Hilary Putnam's
Philosophical Naturalism

Hilary Putnam's Philosophical Naturalism

Making Philosophy Matter for Life

Massimo Dell'Utri

LEXINGTON BOOKS

Lanham • Boulder • New York • London

Published by Lexington Books
An imprint of The Rowman & Littlefield Publishing Group, Inc.
4501 Forbes Boulevard, Suite 200, Lanham, Maryland 20706
www.rowman.com

86-90 Paul Street, London EC2A 4NE

British Library Cataloguing in Publication Information Available

Library of Congress Cataloging-in-Publication Data

Names: Dell'Utri, Massimo, author.
Title: Hilary Putnam's philosophical naturalism : making philosophy matter for life /
    Massimo Dell'Utri.
Description: Lanham : The Rowman & Littlefield Publishing Group, Inc., [2024]
    I Includes bibliographical references and index. I Summary: "Hilary Putnam's
    Philosophical Naturalism: Making Philosophy Matter for Life offers a faithful
    illustration of the trajectory of Putnam's thought to show how, despite the shifts
    in opinion on issues of central philosophical importance, his thought reveals a
    systematic backbone and strong continuities"—Provided by publisher.
Identifiers: LCCN 2024002155 (print) I LCCN 2024002156 (ebook) I
    ISBN 9781666912319 (cloth) I ISBN 9781666912326 (epub)
Subjects: LCSH: Putnam, Hilary. I Naturalism. I Science—Moral and ethical aspects.
Classification: LCC B945.P874 D446 2024 (print) I LCC B945.P874 (ebook) I
    DDC 146—dc23/eng/20240315
LC record available at https://lccn.loc.gov/2024002155
LC ebook record available at https://lccn.loc.gov/2024002156

To Bellocci
without whom…

Contents

Chronology of Life and Work ix

Introduction 1

Chapter One Necessary and A Priori: A Revisitation 9

Chapter Two The Nature of Mathematics 25

Chapter Three What Is the Human Mind 51

Chapter Four The Causal Conception of Meaning 63

Chapter Five The Philosophy of Science 81

Chapter Six The Faces of Realism and Truth 93

Chapter Seven Natural Realism 125

Chapter Eight Moral Philosophy and Philosophy of Religion 161

Chapter Nine Learning from Pragmatism 189

Epilogue: The End of a Journey 209

Bibliography 213

Index 233

About the Author 241

Chronology of Life and Work

1926

Hilary Putnam is born July 31 in Chicago, Illinois, to Samuel Putnam (born Oct. 10, 1888 into a Calvinist family of at least three generations) and Riva Lillian Sampson (born Nov. 29, 1883 into a Jewish family, to parents born in Kaunas in what was the Grand Duchy of Lithuania, later part of the Russian Empire). Samuel Putnam has employment as a journalist and later becomes a translator and writer. He obtains a contract from a publisher to translate the entire works of François Rabelais, and moves with his family to France when Hilary is about 6 months old. Among the people who visit the family "to see the baby" are Ford Madox Ford and Luigi Pirandello. Little Hilary's first language is, therefore, French.

1933

The family returns to the United States, settling in Philadelphia, Pennsylvania.

1944

Graduates from Central High School in Philadelphia. Begins taking courses in German Literature, Linguistic Analysis (the latter also taken by Noam Chomsky) and Philosophy at the University of Pennsylvania.

1948

He graduates from the University of Pennsylvania. His major subject is Philosophy, although he fulfills all the requirements for a major in the other two subjects he took as well. In September he moves to Cambridge, Massachusetts, where at Harvard he takes three courses in Philosophy, two courses in

Mathematical Logic (one with Willard Van Orman Quine and one with Hao Wang), a course in Linear Algebra, and an advanced course in Mathematics: Ideal Theory. In November he marries Erna Diesendruck.

1949

He continues his studies in Philosophy at the University of California at Los Angeles. He has Stanley Cavell as a colleague. Studies with Hans Reichenbach, celebrated exponent of Logical Positivism, "unquestionably the best teacher I ever had" (Putnam 2015b: 17).

1950

On January 15, his father dies.

1951

He obtaines his Ph.D in Philosophy by discussing a thesis entitled *The Meaning of the Concept of Probability in Application to Finite Sequences*, written under the direction of Reichenbach. In the thesis he defends the rule of induction and proves a theorem on the frequentist interpretation of the probability, advocating a verificationist position close to that of his supervisor.

1952

He is assistant professor at Northwestern University, Evanston, Illinois.

1953

He is assistant professor at Princeton, New Jersey. "My eight years on the Princeton faculty were the years of my real *Bildung*" (Putnam 2015b: 26): indeed, he begins to develop innovative and original ideas in philosophy of mind, philosophy of science, philosophy of mathematics and in mathematical logic, thanks to constant and extremely intense work both as a philosopher and as a mathematician. Of great stimulus is his intellectual exchange with Georg Kreisel and Martin Davis, as far as mathematics is concerned, and Rudolf Carnap, another important exponent of Logical Positivism, regarding philosophy.

1955

Daughter Erika is born.

1957

He spends a semester in Minneapolis at the Minnesota Center for the Philosophy of Science, where he meets Herbert Feigl and Wilfrid Sellars, in opposition to whose theses he writes the first draft of "The Analytic and the Synthetic," moving away from the initial verificationist position. He deepens knowledge of quantum mechanics. At the same time he works with Martin Davis on the solution of Hilbert's so-called "tenth problem," that of finding a decision method for the diophantine equations.

1958

Together with Davis he publishes "Reductions of Hilbert's Tenth Problem," their first article on the question. Fruit of the collaboration between the two will be not only a negative solution to David Hilbert's Tenth Problem (i.e., a demonstration that the method of decision hypothesized by Hilbert does not exist), but also the Davis-Putnam algorithm for the satisfiability of formulas of propositional calculus in conjunctive normal form (an algorithm later improved together with others and now known as DPLL, from Davis-Putnam-Logemann-Loveland).

1959

Working on the article "Minds and Machines," he begins to sketch functionalism in philosophy of mind, about which he later states "I was a full-time mathematical logician as well as a philosopher and thus it is not surprising that the idea that the 'mental states' of robots and possibly of people as well would be precisely their computational states should occur to me" (Putnam 2015b: 52).

1960

He becomes Associate Professor at Princeton, both in the Department of Philosophy and in the Department of Mathematics. He obtains a sabbatical year, which he spends in Oxford and Paris. He is invited to lecture at the International Congress in Logic, Methodology and Philosophy of Science where, on August 27, he presents "What Theories Are Not" and meets Ruth Anna Jacobs. He devotes the rest of the year to mathematical research; in particular, to the attempt to prove that the axiom of choice cannot be proved in the Zermelo-Fraenkel standard set theory.

1961

Moves to Cambridge, Massachusetts, accepting an invitation to create a Department of Philosophy at the Massachusetts Institute of Technology, becoming its director. In the *Annals of Mathematics* appears the article "The Decision Problem for Exponential Diophantine Equations" (by Putnam, Davis and Julia Robinson), the result of intensive work on Hilbert's tenth problem.

1962

Marries Ruth Anna Jacobs. Publishes "The Analytic and the Synthetic," "the paper in which I found my own philosophical voice" (Putnam 2015b: 35), "It Ain't Necessarily So," "What Theories Are Not."

1963

Son Samuel is born. In comments on a paper by J. J. C. Smart, presented at a conference in Boston, he lays out the rudiments of what will become his semantic externalism, a topic on which he will work in increasing depth. For the purpose of opposing policy government and the war in Vietnam, he becomes actively involved in organizing of a committee of faculty and students, and several protests within the university. Together with several colleagues, he creates the Boston Area Faculty Group on Public Issues, aimed at collecting money from faculty members of U.S. universities and buy full pages of the *New York Times* to advertise anti-war initiatives.

1965

Son Joshua is born. Begins his teaching at Harvard University, which will last until his retirement. Publishes "A Philosopher Looks at Quantum Mechanics," in which he presents an interpretation of quantum mechanics in agreement with scientific realism (which he considered to be equivalent to the negation of operationalism). He becomes the official representative of Harvard faculty within Students for a Democratic Society, the leading anti-Vietnam War organization.

1966

Daughter Polly (Maxima) is born.

1967

He published "Psychological Predicates"—later republished under the title "The Nature of Mental States"—and "The Mental Life of Some Machines," in which functionalism is explicitly defended. He also publishes "Mathematics without Foundations," in which he proposes an original philosophical conception of mathematics. He actively participates in New England Resistance, an organization supporting young people who refuse to be drafted for an unjust war and transferred to Vietnam.

1969

To make his pacifist militancy more incisive, he joins the Progressive Labor Party, a Maoist-inspired communist group.

1970

Publishes "Is Semantics Possible?"

1971

Publishes *Philosophy of Logic*.

1972

After months of deeply rethinking his relationship with communism, he abandons the Progressive Labor Party, convinced that the recurrent use of the adjective "democratic" made by its leaders is nothing but a farce: "Marxism-Leninism, I came to realize, is not just an intellectual error, but a terrifying sickness of the soul" (Putnam 2015b: 82). However, he will never leave the conviction that philosophers have a responsibility social and political, and not just academic. From then on he limits political activism to the support of Amnesty International.

1975

Publishes "What is Mathematical Truth?," "The Meaning of 'Meaning'," and the first two volumes of his *Philosophical Papers*. As the atheist he has always been, he approaches the Jewish religion (he will celebrate his own Bar Mitzvah in 1994).

1976

In a conference held in Boston entitled "Realism and Reason" repudiates the realist option that since the 1960s had been the background metaphysics to his philosophy, and embraces a verificationist position he calls "internal realism.' He is elected president of the American Philosophical Association (Eastern Division). He is invited by Oxford University to give the John Locke Lectures.

1978

Publishes *Meaning and the Moral Sciences*.

1979

On December 27, his mother dies.

1981

Publishes *Reason, Truth and History*, in the first chapter of which he expounds the famous Brain in a Vat hypothesis (an inspiration for the directors of the movie *Matrix*).

1983

Publishes the third volume of his *Philosophical Papers*.

1988

Publishes *Representation and Reality*, in which he repudiates the functionalist hypothesis in the form he advanced in the early 1960s.

1990

Publishes *Realism with a Human Face*. In his talk at a conference in St. Andrews, Scotland, dedicated to his philosophy (the "Gifford Conference") abandons internal realism and the verificationist semantics associated with it.

1992

Publishes *Renewing Philosophy* and *Pragmatism: An Open Question*.

1994

Publishes *Words and Life*. He is invited by Columbia University to give the "John Dewey Lectures," during which he presents "natural realism," his final metaphysical position. The lectures will be published the same year under the title "Sense, Nonsense, and the Senses: An Inquiry into the Powers of the Human Mind."

1999

Publishes *The Threefold Cord: Mind, Body, and World*.

2000

Leaves Harvard professorship due to age limit and is appointed Professor Emeritus.

2002

Publishes *The Collapse of the Fact/Value Dichotomy and Other Essays*.

2004

Publishes *Ethics without Ontology*.

2007

Participates in the conference organized in Dublin to celebrate his 80 years. From the conference is derived the volume, *Reading Putnam*, released in 2013 and edited by Maria Baghramian.

2008

Publishes *Jewish Philosophy as a Guide to Life: Rosenzweig, Buber, Levinas, Wittgenstein*.

2011

Participates in a conference organized by Harvard University and Brandeis University to celebrate his 85th birthday. From the Kungliga Vetenskapsakademien, the Royal Swedish Academy of Sciences, he is awarded the Rolf Schock Prize in logic and philosophy (the equivalent of the Nobel Prize).

2012

Publishes *Philosophy in an Age of Science: Physics, Mathematics, and Skepticism*.

2015

He is awarded the Nicholas Rescher Prize for Systematic Philosophy by the University of Pittsburgh.

2016

Publishes *Naturalism, Realism, and Normativity*. Dies March 13 in Arlington, Massachusetts.

2017

Pragmatism as a Way of Life: The Lasting Legacy of William James and John Dewey, with Ruth Anna Putnam, is published posthumously.

2022

Philosophy as Dialogue is published posthumously.

Introduction

Hilary Putnam was a unique thinker. Beyond the obviousness of such a statement—every philosopher is unique in possessing a distinctive philosophical profile—Putnam's uniqueness consisted of being an extremely versatile person, intellectually curious, creatively imaginative, humanly sensitive, and honest in acknowledging his own mistakes. Above all, a person actively present in his own time not only from a philosophical point of view, but also from an existential one.

His early education was decisive. A reader of Marx, Kierkegaard, and Freud, and endowed with considerable mathematical ability, upon entering university in 1944 his interests guided him toward courses in German Literature, Linguistic Analysis, and Philosophy; as regards mathematics, he was self-taught at least until 1948, when he decided to take courses in mathematics and algebra at Harvard. But the most important influence he received a few years later from two great logical positivists, Rudolf Carnap and Hans Reichenbach (particularly from the latter, under whose direction he wrote his Ph.D. dissertation), and thus from a type of philosophy in which the sciences—in particular physics—were held in the highest regard. It is to this period that dates Putnam's respect for the natural sciences, his study—frequently renewed in later years—of physics, especially the theory of relativity and quantum mechanics, and the consequent implicit development of a general *naturalistic* perspective—taking "naturalism" to mean the adoption of a metaphysics and epistemology modeled on the natural sciences.[1] But, in Putnam's case, not merely modeled on them. Indeed, his naturalist perspective was never "reductionist" or "eliminativist," always opposing *scientism*—the view that both metaphysics and epistemology should be scientific solely in the sense of the natural sciences:[2]

I take science seriously and [. . .] I regard science as an important part of man's knowledge of reality; but there is a tradition with which I would not wish to be identified, which would say that scientific knowledge is all of man's knowledge. I do not believe that ethical statements are expressions of scientific knowledge; but neither do I agree they are not knowledge at all. The idea that the concepts of truth, falsity, explanation, and even understanding are all concepts which belong exclusively to science seems to me to be a perversion. That Adolf Hitler was a monster seems to me to be a true statement (and even a "description" in any ordinary sense of "description"), but the term "monster" is neither reducible to nor eliminable in favor of "scientific" vocabulary (Putnam 1975a: xiii–xiv).

In addition, a naturalism interpreted in a realist key—the key that enabled Putnam to counter logical positivism critically and in original ways.

The fact is that the study of philosophy, mathematics, and physics was never everything for Putnam: sensitivity to social injustices and engagement in the political sphere were a constant since his first year as an undergraduate. Especially years later when, already a professor, pressing political issues— most conspicuously, the war in Vietnam—could not leave one indifferent, and demanded a firm stance. Hence the commitment to oppose policy government regarding the war, the creation of a committee of faculty and students, the co-founding of the Boston Area Faculty Group on Public Issues, and the organization of several protests within the university. A political activism that involved his own philosophical activity as well:

It seemed to me—and it still seems to me—that it was imperative to try to make a better world, and that philosophy should have its part in trying to bring such a world into being (Putnam 1992g: 349). I think that one of the things I have kept from that period is the idea that philosophy is not, cannot, and should not be simply an academic discipline. I think I owe this to the sixties, which was in many ways a transformative period in my life (Putnam 1994h: 59).

However transformative that period may have been, I do not think it was these events that gave rise to Putnam's awareness that interest in philosophy is not separable from interest in social and civic issues: if anything, they reinforced it. Indeed, I think that such awareness is an expression of an individual's own character, sensibility, and intelligence, and so it was always there. And if sensitivity to social and civic issues amounts to sensitivity to moral and existential issues, then if one is engaged in philosophical reflection it becomes practically inevitable that one broadens one's reflections also to issues of ethics and issues concerning the characteristics of a just and authentically democratic society. It's just that, at that point in his career, the philosophical field within which he had mostly been moving became too cramped: in Putnam's eyes, the area of philosophical research derived from logical positivism and roughly identifiable with the label "analytic philosophy" lacked the conceptual resources necessary to address these issues. Therefore, Putnam broadened the range of concepts and philosophers with whom to engage in dialogue, turning to classical pragmatists and authors traceable—again, loosely—to so-called "continental philosophy:"

> William James used to say that there was no one method for finding truth, not even making predictions or sensory experience. We have to keep having experience and keep discussing it, we have to keep on talking and keep on experiencing. The ethic of democracy, that it is our right to test our political ideas, stops short of denying others the same right. This is connected with a postmodern notion of truth, and it seems to me to be the common element of pragmatism and the New Frankfurt school, one I find very attractive (Putnam 1994h: 63).

Thus began a very important period for Putnam, who not only did not abandon the philosophical concerns that had engaged him from the beginning, but gave them new life thanks both to the new field of study opened up and to a small personal revolution: from the atheist he had been, he became a believer, embracing the Jewish religion around the mid-1970s. A period so decisive that it could be regarded as a watershed—as long as we do not forget the elements of continuity. As has been said,

> From the mid-1980s on, it is not an exaggeration to say that a whole new Hilary Putnam bursts onto the philosophical scene [. . .] His seminars, conversations, essays and books begin regularly to feature names which at most only rarely occur in his writings prior to the mid-1980s (Conant 2022: 36 and 38).

The most immediate effect of the openness to ethics, and to the humanities in general, and especially the effect of the cognitivist approach underlying this openness—an approach dictated by the general realist attitude, with the various modifications to which he subjected it, which Putnam never abandoned—was to broaden the scope of what counts as "reality." Indeed, if we can speak of "knowledge" even in the human sciences, and thus of truth and justification, then this is a clear indication that there is *something* that allows our claims to be *true*—a recognition that leads to the realization of the existence of a plurality of ways of "being true," all subject to the same methodology of testing and retesting that we have in the natural sciences. While the teaching of the classical pragmatists is evident ("I think that the key idea, which Dewey makes very explicit, is that if we assume that there is such a thing as truth in ethics or truth in politics, then it must be subject to the same constraints as scientific truth," Putnam 1994h: 63), it is equally clear that the image of reality is expanded by presenting a plural and multi-level world ("the world has many levels of form, including the level of morally significant human action, and the idea that all of these can be reduced to the level of physics I believe to be a fantasy," Putnam 2008a: 5–6). And the Putnamian naturalist perspective accommodates all this, even giving it a name.

In 1999 John McDowell, distinguishing between a naturalism with a reductivist tendency and a naturalism that grants autonomous space for the human capacity to think and know, called the former "restrictive" and the latter "liberal."[3] A few years later, Mario De Caro and David Macarthur extensively emphasized in a series of writings and anthologies the inescapability of liberal naturalism and brought it close to Putnam's position; in turn, Putnam was happy to borrow this phrase to identify his position.[4]

Thus, Putnam's naturalism is "naturalist" because it rejects any supernatural approach to the study of nature—an approach that admits of entities and modes of knowledge that are not compatible with the framework of the natural sciences—and it is "liberal" because it does not attempt to trace philosophy and the humanities in general back to the successful natural sciences, such as physics. In the terms proposed by Sellars, Putnam claims that "there are not sharply delineated 'scientific images' and 'manifest images,' but forms of human discourse that interpenetrate and depend on one another" (Putnam 2016c: 129). This is a non-scientistic and more inclusive naturalism in that it places the natural sciences, the humanities, and the commonsense perspective on the same level of dignity in light of a broader conception of *experience* that includes our everyday experiences.

> In calling renewed philosophical attention to our everyday lives without reductive or eliminative tendencies, liberal naturalism puts various nonscientific concepts at the center of philosophical attention, for example: persons and their responsibilities and commitments; artifacts including the built environment; the space of first- and second-person relations including rational relations; moral and political values; and (our experience of) artworks and their significance (De Caro & Macarthur 2022b: 3).

It is this liberalized and inclusive perspective that is outlined in this book, following the evolution of Putnam's thought from its emancipation from logical positivism and, partially, from the thought of another leading figure in the philosophical debate of the second half of the last century, Willard Van Orman Quine. During Putnam's formative years, in fact, a radical change was being wrought by Quine's theses, expressed especially in the essay "Two Dogmas of Empiricism" (1951). Quine was highly critical of two cornerstones of logical positivism—the distinction of statements into analytic and synthetic, and reductionism, that is, the idea that every meaningful statement is reducible to a given set of experiences (verifying or falsifying). Therefore, it was inevitable for Putnam to confront Quine as well, and in Chapter 1 we shall see the subtle critical remarks Putnam directed at Quinean arguments.

As mentioned above, it was in opposition to the logical positivists that a realist interpretation of science began to emerge in Putnam's thought in the late 1950s and early 1960s; at the same time, original and profound interpretations of the concepts of the necessary and the a priori were developed. Moreover, Putnam was strongly convinced that meaning is a genuine semantical concept: as he came to declare, "I do not follow Quine [. . .] in his doctrine of the almost total indeterminacy of *reference* (and of *meaning* even when the context is specified)" (Putnam 1986: 301). Accordingly, he proposed a causal elucidation of it, an illustration of which can be found in Chapter 4.

Since, as also mentioned above, simultaneous with his philosophical activity, Putnam was developing in those years intense and fruitful work as a mathematician, it is not surprising that these parallel interests began to interact at some point, giving rise on the one hand to an original philosophical conception of mathematics, *modalism*, and on the other to a theory in philosophy of mind that would influence generations of scholars, *functionalism*. Chapters 2 and 3 respectively are devoted to treating them.

As far as metaphysics is concerned, from the 1970s until the last months of his life, Putnam focused steadily on the most plausible form to be given to his initial realist position and on the most correct way to describe the relationship that connects mind and language to the world. Essentially, reflection on realism marked Putnam's entire career, enriched by additions, rethinking, and confirmation: Chapters 6 and 7 bear witness to this long journey.[5]

Chapters 8 and 9 undertake to show how, according to Putnam, philosophical study has value insofar as it does not separate reflection and life. In his view, as highlighted above,

> philosophers have a double task: to integrate our various views of our world and ourselves [. . .] and to help us find a meaningful orientation in life (Putnam 1989a: 52).

This is in brief the task of *making philosophy matter for life*, a task that reveals how our *practice*, and therefore the *agent's point of view*, takes on an inevitable philosophical centrality.

This is what the reader will find in this book. I may just add that its main aim is to let Putnam speak, as it were, strictly adhering to the texts by incorporating a large number of quotations and precise references. It follows the chronological development of his thought—this is why it starts with the philosophy of science and of mathematics and then addresses his position as a moral and pragmatist philosopher. In so doing, the book aims to offer a faithful illustration of the trajectory of Putnam's thought—not just a critical discussion of it—and to show how, despite the shifts in opinion on issues of central philosophical importance, his thought reveals a systematic backbone and strong continuities.

I said at the outset that Putnam is a unique thinker, and I hope the following chapters will be able to show that. But this very uniqueness poses serious problems for those like me who take on the task of giving an account of it: both because to be truly comprehensive one book is not enough, and because to deal adequately with everything Putnam dealt with would rather require a team embracing a variety of expertises.[6] Nevertheless, I believe that I have succeeded in conveying the Putnamian idea that philosophical analysis "has always been a way of coming to understand both some of the nature and some of the limitations of human reason" (Putnam 1975a: xiv), where the emphasis falls on the word "limitations," since the vast web of human knowledge is bound to contain cognitively impenetrable blind spots and to make the sense of *mystery* an inherent characteristic of us:

> Philosophy is not a subject that eventuates in final solutions, and the discovery that the latest view—no matter if one produced it oneself—*still* does not clear away the mystery is characteristic of the work, when the work is well done (Putnam 1988: xii).

I would like to conclude this Introduction by highlighting this line, first, because I am deeply convinced that it represents our profound human nature (we are unable to *completely* understand ourselves, it seems to me, despite the great advances we have made over the centuries and those we will make in the future); second, because to my knowledge there have been few philosophers broadly adhering to the vast and by now indistinct area called "analytic philosophy" who are willing to say that mystery and the sense of mystery is inherent in us; and third, because the notion of mystery shows at the same time the centrality of the human sciences in our worldview—alongside the natural sciences—and the gist of a liberalized naturalism:

> Should we *regret* the fact that the social sciences cannot realistically hope to resemble physical science? To ask this is to ask if we should regret the fact that we cannot understand ourselves and each other as the physicist understands the harmonic oscillator (nor—what is surely very different—the way God might see us). If we are doomed to have neither a computer's-eye view nor a God's-eye view of ourselves and each other, is that such a terrible fate? We are men and women; and men and women we may be lucky enough to remain. Let us try to preserve our humanity by, among other things, taking a humane view of ourselves and our self-knowledge (Putnam 1978a: 77).

Finally, a few thanks. The people who played a role in various ways in the writing of this book are many, and I shall probably forget some here: for that, I apologize in advance. Certainly, I owe a debt of gratitude to Hilary Putnam, with whom I had several opportunities to discuss and to learn how great philosophical wisdom is never divorced from an admirable human caliber. Just as surely, my debt extends to Mario Alai, Paolo Artuso, Robert Audi, Fabio Bacchini, Maria Baghramian, Carla Bagnoli, Rosa Calcaterra, Stefano Caputo, Sanjit Chakraborty, Roberta Corvi, Richard Davies, Mario De Caro, Juliano do Carmo, Rosaria Egidi, Filippo Ferrari, Pierfrancesco Fiorato, Joe Friggieri, Sossio Giametta, Jonathan Knowles, Antonio Lizzadri, David Macarthur, Giancarlo Marchetti, Sarin Marchetti, Diego Marconi, Sebastiano Moruzzi, Ricardo Navia, Massimo Onofri, Alfredo Paternoster, Carlo Penco, Pietro Perconti, Luigi Perissinotto, Antonio Rainone, Pietro Salis, Filippo Sani, Paolo Valore, Giorgio Volpe, Alberto Voltolini, and Stephen L. White. The criticisms and suggestions of the anonymous referees also proved to be very helpful, and I thank them very much. Special thanks to Rebecca Steinitz and Samuel Putnam. I would also like to thank Deanna Biondi and Jana Hodges-Kluck of Lexington Books, who have shown so much interest in the project and sustained me with great patience and care. The greatest debt, however, is to the dedicatee of this book, my wife Isabella Pesce, for lovingly supporting me over the long span of time that writing the book required.

NOTES

1. On naturalism cf. French, Uehling & Wettstein (1994), Braddon-Mitchell & Nola (2009), do Carmo (2015), Clark (2016), Giladi (2019).

2. Cf.: "The 'naturalism' I reject is, of course, scientistic and reductionist. That, in a *non*-reductionist sense of 'nature,' we and our abilities are part of nature is undeniable" (Putnam 1992g: 403).

3. Here is McDowell a little more in full: "The first approach—a restrictive naturalism—aims to naturalize the concepts of thinking and knowing by forcing the conceptual structure in which they belong into the framework of the realm of law. [The] second approach—a liberal naturalism—does not accept that to reveal thinking and knowing as natural, we need to integrate into the realm of law the frame within which the concepts of thinking and knowing function. All we need is to stress that they are concepts of occurrences and states in our lives" (McDowell 1999: 95).

4. On liberal naturalism cf. De Caro & Macarthur (2004, 2010, 2015, 2022a). On Putnam's liberal naturalism, cf. Putnam (2004b), the articles in De Caro & Floyd (2020), Dell'Utri (2022), De Caro (forthcoming).

5. I would go so far as to claim that anyone who wants to address the question of realism and the most appropriate way of formulating it, cannot help but approach the

ways in which Putnam addressed this question and the debate they provoked. It is something of a must for a contemporary philosopher.

6. For example, one will not find in the book a discussion of his relationship with Wittgenstein (cf. Putnam & Bouveresse 2020), or a discussion of his aesthetic ideas (on this cf. Putnam 1976b; de Gaynesford 2022), or an analysis of his interpretation of quantum mechanics (cf. Maudlin 2022).

Chapter One

Necessary and A Priori

A Revisitation

A RED THREAD

The 1962 paper "The Analytic and the Synthetic" represents a milestone both in the development of Putnam's thought and—I would go so far as to say—in the history of contemporary philosophy itself. For, on the one hand, the conclusions reached in this and a number of other essays from that period (another is "It Ain't Necessarily So," also from 1962) were destined to have an influence on other areas subject to Putnam's careful scrutiny, such as the philosophy of mathematics, of geometry, of quantum mechanics and of language, constituting a red thread that runs through Putnam's thought right up to the works of his maturity; and, on the other hand, the original conceptions concerning the concepts of a priori, necessity, analyticity, and meaning represented a novelty that clearly set Putnam apart from most previous thinkers.

For centuries, philosophers have seen very close relations among the aforementioned concepts, so as to come to conceive—with the logical empiricists of the last century—of a sort of overlap among the first three. According to them, an analytic statement is also necessary and a priori. Putnam's merit is that he offered reasons for untangling these concepts, while at the same time giving them an interpretation that was revolutionary in some respects. He arrived at this interpretation by developing some of the ideas of an illustrious predecessor, a philosopher who greatly influenced his thinking.

In 1951, Willard van Orman Quine published the essay "Two Dogmas of Empiricism" in which he showed, in the first part, the impossibility of rationally justifying the distinction between analytic statements—those that are true or false in virtue of the meaning of the component terms—and synthetic statements—those that are true or false in virtue not only of the meaning of the terms but also of the data of experience. And, in the second part, the impossibility of justifying the "reduction" of the terms of a natural language to the terms of a language of sensory data, and thus the impossibility of justifying a translation of the utterances of, say, English into the empirically meaningful utterances of the language of sensory data. Analyticity and reductionism, Quine concluded, are nothing but dogmas: lacking rational support, they can only be maintained by faith. Due to the overlapping of the concepts of analyticity, necessity, and apriority implemented by the logical empiricists (the main target alluded to in the title of the essay), it is all three concepts that fall under the mallet of Quinean skepticism. Eleven years on, Putnam drew up a general assessment of the plethora of reactions provoked by Quine, presenting a position that accepts many of his conclusions and, at the same time, severely criticizes them: the concepts of analyticity, necessity, and apriority have genuine content, even if it is not exactly that reported by the philosophical tradition. Putnam thus highlighted how simplistic it is to assume that every statement of the language we speak is either analytic or synthetic in nature, and emphasized how, on the contrary, clarifying the existence of a multiplicity of true statements of a very different nature is the most important work a philosopher can do. And this

> not because philosophy is necessarily about language, but because we must become clear about the roles played in our conceptual systems by these diverse kinds of truths before we can get an adequate global view of the world, of thought, of language, or of anything (Putnam 1962b: 41).

THE CONCEPT OF ANALYTICITY

Most of the reactions to the celebrated 1951 Quinean essay are deemed irrelevant by Putnam. Many defenders of the distinction between analytic and synthetic truths have done nothing more than present a series of examples in rebuttal, in the belief that it was sufficient to show concrete cases of analyticity to declare Quine's critique implausible. This, however, Putnam notes, is insufficient: what is needed is to show *why* it is implausible, while providing *reasons* to justify the belief that there are analytic truths. Among the few deserving responses addressed to Quine, Putnam cites that of Paul Grice and Peter Strawson, "who offer *theoretical* reasons for supposing that the analytic-synthetic distinction does in fact exist, even if they do not very satisfactorily delineate that distinction or shed much real light on its nature" (ibid.: 34–35). In particular, their thesis that speakers' *agreement* on the use of certain expressions indicates the presence of some distinction is for Putnam correct and important (cf. Grice and Strawson 1956: 142–143). Note in passing Putnam's inclination to emphasize what speakers actually do in their everyday verbal exchanges: this is an aspect that is destined to constantly characterize his philosophy, a criterion that is of methodological importance when arriving at a decision in controversial cases.

Putnam thus takes Quine's side, but at the same time (as shown by his consonance with part of Grice's and Strawson's position) believes that Quine went too far in asserting that since it is not possible to argue rationally for the existence of analytic truths, *all* truths are synthetic—where the sense in which the word "synthetic" is understood here coincides with the sense of the word "empirical." There are various kinds of statements, and lumping them all together under the label of "empirical" does not seem to do justice to this variety: think for instance of the statements of mathematics, those of logic, those with which we formulate fundamental laws of nature, those of ethics. Moreover, there is an epistemologically relevant consequence of the assumption that all statements are empirical: the consequence is that both the statements traditionally held to be empirical, such as "In the Department's garden there is a rose bush," and the mathematical and logical statements, such as "2 + 2 = 4" and "not (A and (not A))" are all subject to the (sometimes) surprising data of experience. *No* statement would be true or false eternally, so to speak, because the truth-value of *each* can change with the acquisition of new experiences and the advancement of knowledge: even statements of logic so central to our system of knowledge that they have been considered obviously valid principles for centuries can lend themselves to criticism and falsification. Quine himself, who has consistently drawn this consequence from his reasoning, nevertheless sensed the danger, not only epistemological but also existential, inherent in such a consequence: to abandon a logical principle so fundamental as to be practically the basis of all our reasoning would be tantamount to committing a kind of suicide. This is why he felt compelled to point out that, nevertheless, there is in us a "vaguely pragmatic inclination" and a conservative attitude coupled with a "quest for simplicity" (Quine 1951: 46) that compels us to maintain the truth of certain statements—such as those of logic—unaltered. All statements are, indeed, on the same level, but they lie on a *continuum* along which they vary in the *degree* of "protection" from the attacks of experience that we decide to give them. Note: "that *we* decide to give them." This implies that, in spite of Quine, the austere "scientifically objective" framework of his account of human cognitive activity has inevitable gaps through which pass elements that can be characterized as subjective and psychological.

Although initially attracted to the idea of the continuum of statements as Quine understands it (cf. Putnam 1962b: 40), Putnam departs from Quine's interpretation by pointing out that in the continuum there is not only a difference in "degree" (i.e., a different mixture of fact and convention), but also in "type" (i.e., a *qualitative* difference between statements: cf. Putnam 1994d: 251). Let us see how.

First of all, he makes it clear that, in the great variety of statements that we can make, the analytic ones are a small minority, involving only a 100 or so words; and what is more, they are a very trivial and unimportant minority:[1] they are the well-known "A mare is a horse," "All bachelors are unmarried," etc. Nevertheless, although trivial and philosophically unproductive, *pace* Quine these statements are there; indeed, Putnam presents the rudiments of a theory intended to explain their analyticity.

> The idea, in a nutshell, is that there is an exceptionless "law" associated with the noun "bachelor," namely, that someone is a bachelor *if and only if* he has never been married; [. . .] Moreover, this exceptionless law has, in each case, two important characteristics: (1) that no other exceptionless "if and only if" statement is associated with the noun by speakers; and (2) that the exception-less "if and only if" statement in question is a *criterion*, i.e., speakers can and do tell whether or not something is a bachelor by seeing whether or not it is an unmarried man (Putnam 1976a: 89; cf. Putnam 1962b: 65).

The "if and only if" statement associated with the noun "bachelor" is something very close to what philosophers of science call a "law" precisely because it is valid without exception (unless the meaning of the noun is changed): its validity has no statistical character, as is the case with socio-logical laws, for example, but—to the best of our knowledge at the time—is absolute. However, this "if and only if" statement differs from a typical law of nature because it is *unique*: there are no other such statements associated with that noun. This is why it functions as a criterion for whether or not a person is a bachelor.[2]

But why consider the "if and only if" statement mentioned above to be an analytic truth? Putnam gives two reasons, one "negative" and the other "positive." The first is that such a statement is not a synthetic one in the usual sense: it cannot be *refuted* on the basis of isolated experiments, nor can it be *verified* by enumerative induction. In fact, to "verify or confute a statement of the form 'Something is an *A* if and only if it is a *B*' in this way requires that we have *independent* criteria for being an *A* and for being a *B*" (Putnam 1962b: 68). The second reason is that, since the "if and only if" statement in question is the only one associated with the word "bachelor," it has no con-sequences for the language system other than allowing speakers to use a pair of words interchangeably—that is, as *synonyms*. In the absence, therefore, of theoretical grounds for accepting or rejecting the statement, it and all state-ments of the same kind represent *arbitrary fixed points* within our natural lan-guage: "arbitrary" because they are accepted as true by speakers[3] even in the absence of grounds, and "fixed points" because they are immune to revision. And yet, despite such arbitrariness, it is rational to believe in them, given the advantages of having synonyms fixed by analytic statements.[4]

> There they are, the analytic statements: unverifiable in any practical sense, unrefutable in any practical sense, yet we do seem to have them. This must always seem a mystery to one who does not realize the significance of the fact that in any rational way of life there must be certain arbitrary elements (Putnam 1962b: 68).

THE CONCEPTS OF NECESSITY AND A PRIORI

Quine's critique of logical empiricism in the name of an empiricism purged of what he considered philosophically harmful dogmas, and thus in the name of a more genuine empiricism, had among its consequences that of a greater adherence to a cornerstone of American pragmatism: *fallibilism*, that is, the idea that *every* product of the human intellect—from the most complex theory to the simplest statement—can be wrong, despite our beliefs to the contrary, and thus probably to be revised or abandoned. Putnam, with his statement that analytic statements are immune from revision, would seem to repudiate fallibilism, or at least place a definite limit on it. We will examine Putnam's interpretation of fallibilism further below, because in order to fully appreciate it, it is necessary to illustrate his position regarding necessary and a priori statements. As already mentioned, the logical empiricists tended to superimpose the concepts of analyticity, necessity, and apriority, and Quine inherited this approach (cf. Quine 1960: 59). His critique of analyticity, therefore, was also understood by him as a critique of the other two concepts: all three are empty according to Quine. As might be expected, the fact that Putnam grants an adequate, albeit restricted, space to analyticity leads him to also save the concepts of necessity and a priori, but in a highly original way. Let us see how.

The concept of necessity handed down to us by the Western philosophical tradition pertains to metaphysics: it concerns the world and what is in it. What is *necessary* is understood as something that cannot fail to be (whether we know it or not). A necessary statement is thus a statement that retains its truth value in eternity: if it is true, it will always be so, and likewise if it is false. The concept of apriority, on the other hand, pertains to epistemology: it concerns our knowledge, whether actual or possible. What is *a priori* is understood as something that is or can be known independently of experience. An a priori statement is thus a statement whose truth-value (whether true or false) is known to us, or can be known to us, by other than empirical means: it is not by ascertaining facts that we know that truth-value. In "Two Dogmas of Empiricism," Quine presents an argument that leads to the negation of the traditional concepts of necessity and apriority, and Putnam endorses its revolutionary scope with great enthusiasm:

> With Quine, I should like to stress the monolithic character of our conceptual system, the idea of our conceptual system as a massive alliance of beliefs which face the tribunal of experience collectively and not independently, the idea that "when trouble strikes" revisions can, with a very few exceptions, come anywhere (Putnam 1962b: 40).

Every statement, belief or thesis belonging to our conceptual system (the system that represents the knowledge we possess at a certain stage of our collective intellectual development) can be corrected or, if appropriate, abandoned (i.e., declared false): this means that there are no necessary elements in such a system—endowed with a privileged status over others. This is because the Quinean critique of reductionism had shown that it is impossible to bind each "individual" element of the system to its own group of verifying or falsifying evidence, and therefore a given counter-evidence affects all elements in unison: they are all "holistically" responsible to unexpected negative evidence. And there is more. If there is nothing necessary, then nothing can be known independently of experience, that is, a priori. If it did exist, a necessary statement would remain so whatever experience we might have, and the knowledge we might derive from it could therefore be independent of experience; conversely, if experience touches all the elements of our conceptual system, then no statement can be a priori. This is how the Quinean critique of the concept of necessity also invests and destroys the concept of apriority. Any statement we might make is inevitably contingent and a posteriori. In the passage quoted above, we saw that Putnam claims that there are "very few exceptions" to this, and as we now know, these are for him represented by the genuinely analytic statements of our language. However, there are in addition "partial exceptions," and it is this conviction that differentiates his position from that of Quine.

With a strategy destined for widespread success in contemporary philosophy, Putnam invites us to consider a science-fiction case: suppose, he says, that "modern physics has *definitively* come to the conclusion that space is Riemannian" and "let us discuss the status of the statement that *one cannot reach the place from which one came by traveling away from it in a straight line and continuing to move in a constant sense*" (Putnam, 1962a: 239–240). Let us call this statement *S*. Note that the case imagined by Putnam does not concern the abstract space of geometry, but concrete physical space, the space in which we actually move and exist: in his hypothesis, in fact, it is physics that has discovered that space is Riemannian; this is an important aspect for understanding what follows. Bernhard Riemann was a mathematician who in the nineteenth century (along with János Bolyai, Nikolai Lobachevsky, and Carl Gauss) proposed an alternative geometry to Euclidean geometry. Prior to the advent of so-called non-Euclidean geometries, for centuries the axioms and postulates provided by Euclid represented the "laws" that the movements of bodies in space necessarily obeyed—laws that were considered certain, indestructible, and eternal, the falsity of which was literally inconceivable. Given, for example, the definitions of "line" and the property of "being parallel to," it followed that the statements "Two parallel lines extend to infinity without ever meeting" and "If a line is perpendicular to another line, all lines parallel to the first are perpendicular to the second" are indubitable truths—a priori and necessary. This being so, *S* is not only true, but its *status* is that of a necessary and a priori truth.

Now, however, according to the case imagined by Putnam, the inconceivable happens: if space is Riemannian, the Euclidean postulate of parallel lines is false, and therefore *S*, which depends on it, becomes equally false: a line can have its own curvature that—placed in a physical environment—allows one in a finite time to reach the place from which one started, while proceeding in the same direction. Something that had the status of a necessary and a priori statement reveals itself to actually be a contingent and a posteriori statement. Is this the actual *status* of *S*? Not exactly, but before we see what Putnam thinks is the correct moral to be drawn from the science fiction case he proposes, let us briefly examine a possible objection.

Someone might indeed say: Putnam's is only a science fiction example, and nothing can be drawn from it "in reality." Not exactly: even if Putnam proposes a hypothesis, it is a very plausible one. Indeed,

> although modern physics does not *yet* say that space is Riemannian, it does say that our space has variable curvature. This means that if two light rays stay a constant distance apart for a long time and then come closer together after passing the sun, we do not *say* that these two light rays are following curved paths through space, but we say rather that they follow straight paths and that two straight paths may have a constant distance from each other for a long time and then later have a decreasing distance from each other (Putnam 1962a: 242).

And indeed, since the general theory of relativity developed by Alfred Einstein at the beginning of the last century, non-Euclidean geometries have become increasingly plausible. And therein lies the point: it is precisely the presence of an alternative, well-corroborated physical theory to the existing one that can bring about a change in the status of a statement such as *S*. In the absence of such a theory, *S* continues to retain its status as a necessary, a priori statement, and—if it ever occurred to anyone to subject it to scrutiny—in no way could it be proven false. Not, for instance, through experiments. Any experiment that could have been devised to prove the falsity of *S* would have been doomed to failure. The results of an experiment must in fact be properly interpreted, and in general, an interpretation is a function of the best theories that characterize our cognitive system at a given moment: this implies that any interpretation that could have been provided in the pre-Riemannian era, resting on the same theories of which *S* was a consequence, could only lead to its confirmation, and any attempts to refute it would only convince, by their repeated failure, that *S* possesses the character of indubitability and certainty. In a word, of necessity.

So what can we say about the *status* of *S* once the Riemannian nature of space has been established? Putnam's answer emerges after a careful examination of two possible reactions: one is to say that *S* is contingent, the other is to say that it is necessary.

Some logical positivists believed that the conclusion that *S* has changed status is wrong. In fact *S*, they argued, despite what is generally thought, makes use of observational terms, that is, terms referring to empirically observable entities, and therefore is and always has been contingent and a posteriori. Hans Reichenbach, for example, one of the leading exponents of logical positivism, argued that "straight line" is an *operational* expression that must therefore be defined *operationally*, that is, on the basis of actual empirical control operations. By virtue of the operational definition proposed by Reichenbach, the expression "straight line" means "path of a light ray"; if this is the case, "it is clear that the principles of geometry always are and always were synthetic" (Putnam, 1962b: 47) and subject to experiment. And, again, it is clear because *S* is contingent and a posteriori. According to the logical empiricist position, considering it necessary and a priori was nothing more than an "illusion."

The logical positivists also provided a diagnosis of such an illusion: the statement *S* and the statement that *S* is a theorem of Euclidean geometry were mixed up—where according to logical positivists Euclidean geometry was a theory mistakenly believed to be abstract and valid in a realm of abstract mathematical objects in which true statements could only have the character of necessity. This is a confusion because

> that the axioms of Euclid's geometry imply *S* is indeed a necessary truth; but there is all the difference in the world between saying that *S* would be true if space were Euclidean, and saying that *S* is necessarily true (Putnam 1975a, Introduction: ix).

However, Putnam argues, there is something wrong with this analysis. Anyone who believes that some illusion is involved in the transition from one geometry to another must provide a plausible account of how that illusion came to take root,[5] and the appeal to repeated and persistent "confusion" is far from a satisfactory account. Indeed, if educated people thought for centuries that something was impossible and inconceivable, the most reasonable thing to say is that this could happen because that something contradicted certain statements that were deemed "necessary"; and what seemed a necessary truth was precisely that physical (non-abstract) space was Euclidean. Far from any confusion, the status of statement S was held to be that of a necessary and a priori truth.

Proponents of the second possible reaction to the case imagined by Putnam acknowledge that S is necessary and a priori, but offer an erroneous explanation. Indeed, they argue that in the transition from Euclidean to a non-Euclidean geometry the meaning of the relevant terms changes: in our case, the meaning of the expression "straight line" (cf. Grice & Strawson 1956: 157). Thus, if the meaning of one or more words in a statement changes, it is no longer the statement it was before but another one, just as in a given circumstance "The bank is washed away" is a true statement if "bank" means "pile of various materials along a river waterfront" and false if it means "financial establishment that invests money deposited by customers": we have two different statements for all that they are expressed with the same words. Similarly, within the framework of Euclidean geometry S has a certain meaning, while it acquires a completely different meaning within the framework of a non-Euclidean geometry, thus becoming another statement. As a consequence of the Euclidean axioms and postulates, however, S is and remains necessary and a priori: on closer inspection—this reasoning continues—when, as Putnam's case assumes, physics definitively proves that space is Riemannian, Euclidean geometry is only apparently false (cf. Putnam 1962a: 241). And, if it is only apparently false, we can always go back to using it should we find it convenient to do so: hence the conviction that the statement S "would still be a necessary truth if the meanings of the words had been kept unchanged" (Putnam 1962a: 248).

Putnam immediately notes that the idea that in the transition from one theory to the next the meanings of the central words (the so-called "theoretical terms") change does not account for everything that actually happens in the usual cognitive dynamic. As the peculiar use of the adjectives "true" and "false" would seem to indicate, the conception of such dynamics underlying the reaction just seen would seem to be vitiated by a precise "metaphysical" hypothesis: the hypothesis that in the world there not only exist straight lines as described by non-Euclidean geometries, but also straight lines that "fill out a Euclidean space" (Putnam 1962a: 242). Both types of objects would exist, and thus the transition from Euclidean to a non-Euclidean geometry would merely apply theoretical terms to new referents. Yet,

Insofar as the terms "place," "path," and "straight line" have any application at all in physical space, they still have the application that they always had; something that was literally inconceivable has turned out to be true (Putnam 1962a: 242).

Apart from the theoretical inconvenience of hypothesizing such an over-populated reality, in general the metaphysics implicit in the second reaction seems hardly in harmony with the dynamics actually at play when in any field of knowledge, and particularly in science, one theory is supplanted by a later theory that better explains the phenomena under scrutiny. Such implicit metaphysics has a typically idealistic flavor: the determination of the meaning of terms, including the determination of their referents, is considered a question that can only be decided within individual theories. Consequently, what counts as "the world" can be deduced from what these theories establish. This is why the meaning and referent of a term, say "straight line," is seen to change as the theory changes: according to this approach, we from time to time apply the term to the type of objects established by the theory on which we decide to base ourselves. Precisely this is contested by Putnam. Is this really how, he asks, scientists behave when they formulate and adopt a new theory—one that they judge to be better in terms of explanatory and predictive power? Actually, it is not reasonable to think that the new theory postulates "new" objects to be placed alongside the "old," new objects to which the usual terms are to be applied,

because, from the standpoint of the new theory, there are not and never were any objects which could plausibly have been the referents of the words in question (Putnam 1975b, Introduction: xv),

even if earlier theorists believed so.

What the new theory does is to offer a more adequate description of the behavior of the same objects to which the terms used by the earlier theory referred. That being the case, the application of the terms does not change: "straight line" retains its referent in the transition from Euclidean to Riemannian geometry; only that what one thinks of as a straight line is now more correct. There is therefore something constant in the transition from one theory to another, namely the world: according to Putnam's realist metaphysics, what changes are not the meanings (or rather the denotations) of terms, but the beliefs about the objects to which those terms refer.[6] We continue to talk about the same objects, but now the new theory provides a more appropriate conceptual apparatus, and it is precisely because of this that we see something that was previously considered "inconceivable": it was so because we lacked the appropriate concepts to conceive it.

What, then, is the moral to be drawn from the science-fiction case imagined by Putnam? What does he think is the actual status of *S*? As can be seen from the following passage, he takes a position that we might consider intermediate between those motivating the two reactions just seen:

> Euclidean geometry *as a theory of physical space* was always a synthetic theory, a theory about the world, but it had the strongest possible kind of paradigm status prior to the elaboration of the alternative paradigm (Putnam 1975a, Introduction: x, italics mine).

A synthetic theory can be undermined on the grounds of experience: observations of natural phenomena, laboratory experiments, and the like. Euclidean geometry, Putnam tells us, is a synthetic theory, but as such very *sui generis*: before the advent of non-Euclidean geometries, no empirical evidence could have demonstrated its inadequacy. It served as a "paradigm," a conceptual background against which alone it was possible to observe, explain, and predict phenomena occurring in physical space. No experiment could prove its inadequacy, because every experiment that one was able to devise inevitably rested on assumptions deduced from it and, therefore, presupposed its validity. And it was precisely because it seemed untouchable by any experiment that it was regarded as a typical example of a necessary and a priori theory. The situation changes radically when alternative and valid theories are presented, the only ones capable of providing theoretical premises on the basis of which experiments aimed at refuting Euclidean geometry can be formulated: its character as a synthetic theory then becomes clear.

What, then, is to be said of statement *S*, a statement derived from a theory that enjoyed "paradigmatic" status until the advent of one or more alternative theories? Putnam's answer is that *S* has the status of a necessary and a priori truth as long as there are no alternatives to the paradigm on which it rests, and the status of a contingent and a posteriori statement when these alternatives appear and prove valid. For statements of this kind he coined the name "*quasi*-necessary relative to a conceptual scheme."[7] The change in status of statement *S* (and with it a *host* of similar statements) is something that concerns "us," our knowledge and the development of our methods for investigating reality. In short, it concerns our attitude toward *S*. It is not that *S* was necessary and a priori before and later became contingent and a posteriori: it has always had the latter status. The fact is that we can only say it is contingent after a certain theory has been elaborated and then corroborated: before the advent of such a theory *S* could not be undermined by any experiment that could be devised, no matter how ingenious. The notion of "quasi-necessity relative to a conceptual scheme" expresses both the status of a statement and our attitude toward that status, thereby denying the notion of necessity in the sense of "eternally necessary." Despite what has been taught in the philosophical tradition, this latter notion would seem to be entirely without content.

"It would seem": but for Putnam, is it really without content or not? Before we look at his actual position on the matter, we must clarify what that "*host* of statements similar" to statement *S* I mentioned a few lines above consists of.

These are the most important building blocks of our conceptual system: for example, the principles of mathematics and logic, the laws of natural science, and the statements that serve as framework principles within the system, such as the thesis that there is a past or the thesis that a person can only be said to possess a certain knowledge if that person is in possession of an adequate justification. As we have seen happen with the laws of geometry, these statements possess

> the characteristic of being so central that they are employed as auxiliaries to make predictions in an overwhelming number of experiments, without themselves being jeopardized by any possible experimental results (Putnam 1962b: 48).

Their existence shows that the attempt to consider our language as consisting solely of statements that are either analytic or synthetic constitutes an undue impoverishment. We obtain a more faithful description if, alongside the small group of genuinely analytic statements and the group of genuinely empirical statements (those susceptible to being corroborated or refuted on the basis of empirical observations), we contemplate the presence of quasi-necessary statements. For the sake of methodological convenience, Putnam proposes to reserve the label "synthetic" for the latter, that is, statements that are indeed revisable in principle (cf. Putnam 1962b: 59), but not on the basis of the results of experience, unless these results are interpreted by means of an alternative theory to the current one. Until such a theory appears and is validated, quasi-necessary statements cannot be touched by experience, and are therefore "non-empirical," "not a posteriori." In a word, they are a priori. This, then, is how, a couple of centuries after Kant, one can speak, according to Putnam, of *synthetic a priori* statements.[8]

SOPHISTICATED FALLIBILISM

It is interesting to note how Putnam's revisitation of the concept of necessity narrows its difference from the concept of a priori. We said above that the concept of necessity is traditionally considered a metaphysical concept, whereas the concept of a priori is an epistemic concept. Well, the fact that a quasi-necessary statement concerns us, that is, what we know at a given historical moment, indicates that the content of the notion of quasi-necessity is in part provided by the knowledge we have been able to elaborate at a given stage of our cultural evolution: the discourse on necessity thus takes on a clear epistemic valence, interweaving the necessary with the a priori. Because of this entanglement, the two concepts share the same fate: must we therefore conclude that for Putnam, the two old concepts must be completely eliminated and replaced by their updated versions? Not quite. Consider how he analyses the concept of a priori.

Following in Quine's footsteps, he identifies apriority with "unrevisability": if a statement is a priori, then it is incorrigible, irrefutable, untouchable, unrevisable. However, it is possible to distinguish at least two interpretations of the concept of unrevisability: one *behaviorist* and one *epistemic*. The first is the more radical: an unrevisable statement is a statement that we would never give up; anything can happen, but between our behavioral dispositions and our actual behaviors there will never be one to eliminate the statement in question.[9] According to the second interpretation, an unrevisable statement is a statement that it is not "rational" to abandon. This is the Putnamian interpretation. Given the canons of rationality in force at a certain stage in the development of our conceptual scheme, there are inevitably statements that it would be completely irrational to abandon or correct: as we know, these are the quasi-necessary statements. However, *pace* Kant, our rationality and the canons it helps to formulate are not something given once and for all, but evolve and change over time: what is rational to accept in a given period, may not be rational in a later period. Therefore, in the same way that we can speak of necessity relative to a stage of development of our conceptual scheme, so we can speak of "contextual a priori," where the context is here represented by a particular moment in the historical development of that scheme. At such moments, to abandon a "fundamental" statement—the postulate of parallel straight lines is a good example—is contrary to the canons of rationality in force in the scheme, and therefore, although contextually, that statement is a priori. But some care is called for here. Does this mean that—as suggested by the celebrated Quinean conclusion that "no statement is immune to revision" (Quine 1951: 43)—for every statement there are circumstances in which it would be rational to abandon it? Does it mean in other words that all statements are falsifiable, as fallibilism would seem to claim? And, conversely, does this mean that the "absolute" a priori, according to which it could *never* be rational to abandon a given statement, is devoid of content? Not at all.

According to Putnam, the sphere of a priori truths in the absolute sense is enormously thinned out, but it is not nullified: there is in fact at least one a priori truth in the absolute sense, and therefore one truth that it would never be rational to abandon no matter how many changes the notion of rationality may go through. Examples are "Not every statement is both true and false" and "Not every statement is true" (the former appears in Putnam 1978d: 101, the latter in Putnam 1981: 83). Admittedly, these are trivial cases (cf. Putnam 1981: 84), and yet we could not abandon these truths by accepting their negation because otherwise we would be forced to state, respectively, that every statement is both true and false, and that every statement is true, losing any basis for qualifying even our most usual reasoning as rational.

This is enough to preserve the traditional absolute concept of the a priori; but, by virtue of the interweaving we mentioned, it is also enough to preserve the traditional absolute concept of necessity. The moral is clear: to try to understand the metaphysics and epistemology underlying our overall conceptual scheme in terms of a single notion of a priori and a single notion of necessity is "a serious mistake" (Putnam 1978d: 100): a philosopher can only work with more than one interpretation of both concepts. And in fact, with regard to the concept of necessity, while on the one hand Putnam borrows the concept of metaphysically necessity from Saul Kripke by making it interact with that of the epistemically contingent (cf. Chapter 4), on the other hand he offers a much less metaphysically connoted account of necessity: it is not that the validity of certain statements, those of logic for example, is inscribed for all eternity in the constitution of human thought by virtue of the special metaphysical nature of this very thought, so as to render those statements eternally unrevisable. More simply, "it is that the question 'Are they revisable?' is one which we have not yet succeeded in giving a sense" (Putnam 1994d: 256). But again we need to be careful: it is not because of our personal shortcomings that we fail to do so. The fact is that we just cannot do it because we do not possess the slightest basis for 'describing' circumstances that would lead us to revise the truth-value of those statements, and it is precisely this lack of sense that makes the idea that such statements can be refuted "inconceivable." In cases such as these, we cannot say either that they are irrefutable (it is the general fallibilist view that forbids us to do so), or that they are refutable. Conversely, to insist with Quine that all statements must be falsifiable only makes falsifiability the "*third* (or is it a fourth by now?) *dogma of empiricism*" (Putnam 1994d: 258, my italics).

To leave it to Putnam:

> certainly there is no metaphysical guarantee available that something that will strike us as completely analogous to what happened in the case of geometry will *never* happen in the field of logic. To point this out is the right way to be a "fallibilist" with respect to logical laws. But to express fallibilism "positively," by saying "The laws of logic may turn out to be wrong," is a mistake: for we have no more succeeded in giving *those* words a sense now than the pre-Euclideans [. . .] had succeeded in giving a sense to "a plane triangle may have two right angles as base angles" (Putnam 1992i: 375–376).

It is therefore in the name of a sophisticated fallibilism that Putnam concludes that,

> [s]ince our notions of rationality and of rational revisability are the product of our all too limited experience and all too fallible biology, it is to be expected that even principles we regard as "a priori," or "conceptual," or whatever, will from time to time turn out to need revision in the light of unexpected experiences or unanticipated theoretical innovations. Such revision cannot be unlimited: otherwise we would no longer have a concept of anything we could call *rationality*; but the limits are not in general possible for us to state (Putnam 1981: 83–84).

NOTES

1. See for example statements such as the following: "I think that there is an analytic-synthetic distinction, but a rather trivial one," "The analytic-synthetic distinction [. . .] is of overwhelming unimportance" (Putnam 1962a: 36 and 41). In this regard Charles Travis states: "Putnam saw himself as aligned with Quine *to an extent*, but also diverging from him to an extent. The alignment consisted in attacking a certain conception of analyticity which yielded too rich a lode for philosophers to mine. The divergence was in insisting that there *was* an analytic-synthetic distinction, though a pretty boring one. The *distinction* may have been banal, but the divergence was not. What Quine insisted was that there was no sense to be made of the idea of analyticity. His case rested on reductionist assumptions which Putnam never accepted" (Travis 2020: 463).

2. Note that the uniqueness of the statement "if and only if" linked to "bachelor" (and its function as a criterion) makes the two expressions "bachelor" and "unmarried man" *synonymous*. After all, the close connection between analyticity and synonymy is at the heart of Quine's critical argument.

3. Recall Grice and Strawson's thesis mentioned above.

4. The advantages cited by Putnam are *brevity* (the possibility of using the short "bachelor" instead of the longer "adult unmarried man") and *intelligibility* (speakers can understand each other better because the existence of fixed points allows them to predict in advance some usages of other speakers): cf. Putnam (1962a): 56.

5. "If it was just an illusion that the statements of Euclidean geometry were necessary truths—necessary truths about the space in which bodies move—then philosophers of science owe us a plausible explanation of this illusion" (Putnam 1975a: ix).

6. Cf. Putnam (1962b): 50. As we will specify later when dealing with the causal conception of meaning, in the course of theoretical dynamics the meanings of natural kind terms or physical magnitude terms change with regard to the stereotype but not with regard to the referent, which is the part Putnam considers most important in the meaning of terms of this type.

7. Cf. Putnam (1994d): 251. This name corrects the analogous "necessary relative to a body of knowledge" of thirty years earlier. A useful discussion of Putnam's notion of quasi-necessity can be found in Narboux (2020) and Travis (2020).

8. Cf.: "the revisability of the laws of Euclid's geometry, or the laws of classical logic, does not make them mere 'empirical' statements" (Putnam 1978a: 138).

9. With some caution, Putnam attributes this interpretation to Quine: cf. Putnam (1978d): 98. Of the two, it is the more radical interpretation because it assumes that it is possible to firmly support a statement (or, in the case of a priori negation, to abandon it) regardless of any consideration about the rationality of that choice.

Chapter Two

The Nature of Mathematics

QUASI-EMPIRICAL MATHEMATICS

As one might have already guessed, the critique of the traditional concepts of a priori and necessity, and their replacement by correlative concepts that are not absolute but relative to the cultural-historical context, could not fail to have a reverberation on Putnam's position in the philosophy of mathematics and logic, since the true statements of both disciplines have for centuries been referred to as models of statements that are knowable independently of any experience, the negation of which is rationally inconceivable. The result of this reverberation is a thinning of the dividing line between mathematics and empirical science: "in mathematics too there is an interplay of postulation, quasi-empirical testing, and conceptual revolution leading to the formation of contextually *a priori* paradigms" (Putnam 1975a, Introduction: xi). One need only look at the actual practice of mathematicians, Putnam argues, to realize that "*we* have been using quasi-empirical and even empirical methods in mathematics all along—we, us humans, right here on earth!" (Putnam 1975f: 64) methods closer to those used by natural scientists than to the demonstrative method traditionally considered typical of mathematicians.

Consider, for example, the introduction of real numbers or the fundamental postulate on which analytical geometry rests, the postulate that there is a one-to-one correspondence between points on a line and real numbers that preserves their order. The ancient Greeks lacked the necessary mathematical experience and concepts to generalize the notion of number, and could therefore not be in a position to offer a rigorous justification for such an introduction. The same essentially applies to Descartes, who merely "postulated" the existence of a number in correspondence with each interval on the line. Yet, "once he had shown how great the 'pay off' of the correspondence postulate

25

was, not only in pure mathematics but also in mechanics, there was not the slightest question of abandoning either the correspondence postulate or these generalized numbers, the 'real' numbers" (Putnam 1975f: 64–65). Herein lies the crux of the matter: what proved decisive in the complete acceptance of the postulate and the real numbers was their *fertility* in both physics and mathematics, their *success* in facilitating calculations and conclusions, and not an indisputable "proof" of their validity.

Of such examples, Putnam continues, the history of mathematics is full. Consider the introduction of the methods of differential and integral calculus by Isaac Newton and Gottfried Wilhelm von Leibniz, or the introduction of the axiom of choice by Ernst Zermelo, or the way in which Leonhard Euler discovered that the sum of the series $1/n^2$ is $\pi^2/6$, or—again—the fact that several mathematicians are convinced, entirely irrespective of mathematical proofs, that there are infinitely many twin primes, that is, infinite pairs n, $n + 2$ of prime numbers (such as 5 and 7 or 11 and 13). In all these cases, mathematical statements are accepted in the complete absence of rigorous and incontrovertible proofs of them—the typical proofs that, starting from axioms accepted as true, carry this truth to the conclusions through the application of rules of inference considered universally valid. Instead of such proofs, we have here arguments that show a very close resemblance to the hypothetico-deductive arguments of the empirical sciences, where certain intuitively plausible ideas lead to results that are subjected to "empirical" scrutiny, and where the positive results of such scrutiny reinforce confidence in those ideas. As can be seen, similarly to what happens in the natural sciences, an important role is played by a subjective and exquisitely human element: *intuition*. Granted, intuition "is a *fallible* guide—that is what Francis Bacon taught us—but a fallible guide is still better than no guide at all. If our intuition were totally untrustworthy, we would never think of a correct or approximately correct theory to test in the first place" (Putnam 1975f: 67). Intuition precisely gives us a glimpse of what might be useful for scientific progress, leading us to introduce new axioms, new theorems or new "objects" on an *experimental* basis:

> Mathematics is not an experimental science; that is the first thing that every philosopher learns. Yet the adoption of the axiom of choice as a new mathematical paradigm *was* an experiment, even if the experiment was not performed by men in white coats in a laboratory (Putnam 1975a, Introduction: xi).

Experiments—or "quasi-experiments"—of this kind represent, when successful, not a proof but a "verification" or "confirmation" of the statement in question. And, like all verifications, such quasi-empirical verification is relative and not absolute, in the sense that what is verified now may turn out to be false at a later quasi-empirical check. This is why mathematics—and in this respect also logic—cannot be considered an a priori discipline in the classical sense: nothing excludes that its fundamental statements can become false.

However, it should be emphasized, Putnam does not intend to consider mathematics (and logic) an empirical science, he does not intend to abolish what is an obvious and important distinction between disciplines.[1] He does not, for example, wish to minimize the importance in mathematics of proofs; on the contrary, "proof will continue to be the primary method of mathematical verification" (Putnam 1975f: 76), and quasi-empirical proof and inference are simply to be regarded as complementary. Mathematics is obviously more a priori than physics, but this does not mean that the fundamental statements of mathematics are completely safe from experience and, therefore, immune to revision.

> We will be justified in accepting classical propositional calculus or Peano number theory not because the relevant statements are "unrevisable in principle" but because a great deal of science presupposes these statements and because no real alternative is in the field (Putnam 1967a: 51).

As in the above-mentioned case of geometry, the fact is that to refute them requires a far-reaching conceptual revolution that changes the background within which those statements are placed and makes a refuting quasi-experiment intelligible. In the absence of an alternative background it remains permissible to consider those statements a priori—*relatively a priori*, or (since they are not common empirical statements) *synthetic a priori*.

But the effects of Putnam's peculiar conception of the a priori in the philosophy of mathematics do not stop there. If mathematical knowledge resembles empirical knowledge insofar as the success of ideas in both constitutes a criterion of their truth, and both are not absolute but revisable, it follows that their objectivity is explained along the same coordinates: both concern a reality independent of the human mind—they concern *facts of the matter*, that is, data independent of human subjectivity.[2] In particular, this leads to a critique against two important positions in the philosophy of mathematics: on the one hand, the form of *nominalism* generally upheld by logical positivists—namely that the principles of mathematics are nothing more than conventions or linguistic rules and therefore do not refer to any reality—and, on the other hand, *Platonism*, which, concerning reality, ends up explaining "too much" and reveals itself to be prey to a widespread but incorrect idea: that objectivity cannot be explained except by reference to objects. Platonism is thus for Putnam not only implausible from an epistemological point of view—because it assumes that we have a priori knowledge of special eternal objects—but also from a metaphysical point of view—because it postulates the existence of these eternal objects themselves. Yet, surprisingly, while the condemnation of the logical positivist position is without appeal, Putnam reserves a different attitude for the Platonist conception of mathematics (the interpretation of mathematics as a theory of sets, conceived as abstract objects), an attitude consisting in considering Platonism and Putnam's own conception of mathematics (*modalism*) as "equivalent descriptions." Since such an attitude—applied in the most diverse contexts—is a constant feature of his thought, let us examine it in detail before turning to Putnam's interpretation of mathematics.

CONCEPTUAL RELATIVITY AND
EQUIVALENT DESCRIPTIONS

One of the keys on which Putnam insisted throughout his career is the phenomenon of "conceptual relativity." In fact, the resolute insistence on conceptual relativity from the mid-1980s onward is what marked the transition from a *pars destruens* to a *pars construens* in his critique of the so-called "metaphysical realism"[3] (which we will address in Chapter 6), where the *pars destruens* is represented by the famous model-theoretic argument and the brain-in-a-vat argument. The phenomenon of conceptual relativity is given by the "cognitive equivalence" of statements, theories, or conceptual systems that, taken literally, are incompatible: for example, two statements that say two different things about the same portion of the world and yet are both true. How can something that, at first glance, is so counterintuitive happen? (Small hint: it all revolves around the just-mentioned locution "taken literally.")[4]

Examples of Equivalent Descriptions

To begin with, let us see which cases Putnam is referring to, and let us start with a particularly apt example: Einstein's special, or restricted, theory of relativity of 1905–07, a theory in which the notion of equivalence is central.[5]

Let us consider two observers, O and O', engaged in the description of the same event. As is well known, any event is mathematically identifiable by means of four coordinates: three spatial, which define the place, and one temporal, which defines the instant in which the event occurred. Suppose that O measures the coordinates of the observed event on the axes of a Cartesian tern and a clock (which from a mathematical point of view is the fourth axis, that of time), and that the result of his measurements is x, y, z, t. Suppose also that O' makes its measurements on other axes, the result being x', y', z', t'. Let us finally assume that the two coordinate systems are inertial, that is, they have no accelerations of any kind, and are therefore either at rest relative to each other or in motion with constant relative velocity. We may thus choose to consider the system of O at rest and that of O' in motion, or vice versa. Be that as it may, the passage from considering the first system as being at rest to considering the second as being at rest corresponds, from a mathematical point of view, to a transformation of the coordinate system, that is, a translation of the event data obtained within the first system into the event data obtained within the other system. In classical physics, this transformation—which is called "Galilean"—entails two consequences: the first is that, provided that the two observers have previously tuned their clocks and that these are physically identical, we have it that for both the measured event occurs at the same instant; furthermore, we have that two events that occur simultaneously for O occur simultaneously for O'. The second consequence is that, given any two points, their distance is the same for both O and O'.

According to the special theory of relativity, however, the transition from the system of O at rest to O' in motion, or vice versa, is not correctly described by a Galilean transformation. Correct predictions of the coordinates of an event in terms of the system of O made from the coordinates of the system of O' are given by a different mathematical transformation, the *Lorentz transformation*. And this leads to conclusions that are radically at odds with the two postulates just seen from classical physics. The measurements of the same event taken within each system will, in fact, record a certain time interval, and two events that appear simultaneous in the first system will no longer be so in the second. Similarly, the distance between any two points will vary from one system to the other: the length of a segment measured on the co-ordinates of the system at rest of O' will be shorter when evaluated from O when it is in motion with respect to O'. The two observers will therefore formulate different descriptions of the same events, and these descriptions are both true—something that at first glance appears patently contradictory.

> Description (A) says two events X and Y (say, an explosion on the moon and an explosion on Mars) happened simultaneously and description (B) says X happened *before* Y. How can two such flatly contradictory accounts *both* be true? (Putnam 1978b: 34)

Note that, although it plays a central role therein, the notion of equivalence is not the exclusive preserve of the special theory of relativity. Many of us are familiar, for example, with the curious situation in quantum mechanics whereby the same phenomenon can be described by two equally valid theories, one based on the existence of particles, the other on the existence of waves: an electron passing through a certain point in spacetime is described by the former as a particle endowed with a well-defined momentum and an indeterminate position, and by the latter as a wave endowed with a particular length. Both theories are legitimized to the same degree by the phenomena under study, in the sense that the facts do not exclude one in favor of the other; yet the two theories state things that are incompatible with each other—at least at first glance.

Nor would it be correct to say that the phenomenon of equivalent descriptions only concerns the sphere of physics, because on closer inspection it is quite pervasive: it is in fact present in all those cases in which it is possible to choose between equivalent alternatives. For example,

> the choice, in different context, of including or not including mereological sums in one's ontology; [. . .] the choice, in formalized geometry, of taking points to be individuals or taking them to be convergent sequences of spheres (or of other solids—not to mention the various different ways in which the notion of "convergent sequence" has been formalized!); [. . .] the choice, in a certain portion of classical electrodynamics between taking the action between charged particles to be mediated by "fields" or by "point-source retarded potentials"; and [. . .] the choice, in mathematical logic, between taking sets to be characteristic functions or taking them to be primitive objects and taking functions to be sets of ordered pairs (Putnam 2001: 432–433).

In all these cases, nothing dictates that one of the two alternatives be chosen instead of the other: no mathematician, in the latter case, "would suppose that this 'ontological difference' had the slightest mathematical *or* metaphysical significance" (Putnam 2001: 432). Each theory and statement does justice to the phenomena in exactly the same way as the alternative theory and statement does.[6] In short, we are faced with pairs of elements that, from an epistemological point of view, are on the same plane. But how should the relationship between the two elements of such a pair be considered? Let us first see how, according to Putnam, we should *not* consider it.

TWO WRONG REACTIONS

At first glance, the cognitive equivalence of theories or statements would seem not to constitute an interesting scientific issue, let alone a philosophical one. In fact, a quite natural reaction is to assert that there is, after all, an independent and external fact—a *fact of the matter*—on the basis of which we can unambiguously establish which of two or more equivalent descriptions is true and which false. According to this reaction, the equivalence does not exist at all, and it is only our cognitive limitations that make us assume it. But such a reaction "leads either to scepticism or to a revival of metaphysics of the kind that Kant persuaded us to abandon" (Putnam 1978b: 43): one would arrive at the former if one assumes that the human mind is not and never will be able to know that fact of the matter, while one would arrive at a metaphysics that admits an a priori knowledge of noumenal "things in themselves" in the event that it is argued that knowledge of which description of the world is true is attainable through extra-scientific methods.

Another reaction, on the other hand, is to admit the phenomenon of equivalent descriptions (i.e., to admit that there is no fact of the matter on which to rely in order to settle the question) but to emphasize the obviousness that things can receive multiple descriptions: it is obvious, one might argue, that one can draw a map of Texas using a polar projection or a Mercator projection, but—the objection would continue—Texas is still "there" in its uniqueness and in its imperturbable independence of the language with which the descriptions are formulated. And it is equally obvious that one can answer the question "How much do you weigh?" by employing as a unit of measurement once the kilogram and another time the pound without any inconsistency whatsoever: "the appearance of inconsistency is only an appearance, because the claim that I weigh 160 in pounds is consistent with the claim that I weigh 73 in kilograms" (Searle 1995: 165). It is the object itself—a person in this case—that allows two or more descriptions that are apparently incompatible, but which are not incompatible at all.

However, while it is true that things can be described from several points of view, the fact is that these points of view are generally not cognitively equivalent and are often compatible with each other. As we shall see later, such cases are not cases of conceptual relativity, but cases of what Putnam calls *conceptual pluralism*. This is why this reaction too—we might call it the Searlean reaction—fails to frame the actual philosophical significance of equivalent descriptions, albeit for different reasons from those underlying the previous objection. It leads to a bad metaphysics and a bad epistemology: to an implausible conception of what the world is and to a distorted image of the cognitive relationship between human beings and the world. In order to make us understand why this is an unfaithful image, Putnam likens it to the relationship between cookie cutters and the dough to which they are applied: the metaphysical-epistemological image that emerges is then that of a fixed and invariant world (the *dough*) that is represented by different conceptual schemes (the *cookie cutters*).[7] But what is wrong with such an image?

The answer is that the metaphor of the cookie cutter, with its idea of a world consisting of a fixed set of objects and their properties, presupposes a dichotomy between *fact* and *convention*, that is, it presupposes that it is possible to distinguish precisely what in our cognitive judgments depends on facts (the dough) and what depends on us (the cookie cutters)—on our conventional theoretical choices. And such a dichotomy is for Putnam untenable.

Fact and Convention

That we cannot clearly distinguish between a conceptual scheme and a portion of the world in the same way as we distinguish the cookie cutter from the dough is a thesis that follows from the Quinean denial of the dichotomy between the analytic and the synthetic, a denial that, as we know, Putnam—albeit with an important distinction—accepts. The validity of an analytic statement is the result of a conventional decision, a linguistic rule for example, and if there were genuine analytic statements in epistemological questions it would be all too easy to distinguish the contribution of convention from that of facts. But, with the exception of a few sparse cases that are more of interest to the linguist than the epistemologist, analytic statements do not exist, according to Putnam: far from there being a clear separation, there is an "interpenetration" between fact and convention. He used several examples to illustrate this, but there is one that has remained the most famous—and perhaps the least understood. It runs as follows.

Let us consider an oversimplified world consisting of only three objects. Since Rudolf Carnap—with whom Putnam studied inductive logic in the 1950s—often used such examples, Putnam calls this "a world *à la* Carnap." Of course, if we were to answer the question "How many objects are there in this world?," we would all answer a firm "Three!"; but the question is not so obvious. Suppose the question were put to an advocate of mereology, the calculus of parts and integers invented by the Polish logician Stanislaw Leśniewski: his answer would be a firm "Seven"—or "Eight" in the case of admitting the null object. Mereology in fact contemplates the possibility of 'adding' objects to form a further object: our Polish logician therefore only needs to take a quick look at the world that Carnap describes as consisting of x_1, x_2, x_3 to state that the objects of the world are x_1, x_2, x_3, $x_1 + x_2$, $x_1 + x_3$, $x_2 + x_3$, $x_1 + x_2 + x_3$.

Which of the two descriptions is the correct one? Here, the point is that there is *no fact of the matter* on the basis of which we can say that one description is right and the other wrong. They are both right, because they are "underdetermined" by facts and guided by our free conventional choice: the choice of a way of speaking, of a conceptual framework, of a way of describing a state of affairs.[8] Putnam thus emphasizes how "what is factual and what is conventional is a matter of degree" (Putnam 1988: 113), a mixture in which the proportions between the two elements vary from case to case. As we have seen in examining the analytic-synthetic question,

> there is continuum stretching from choices which, by our present lights, are just choices of a way of talking to questions of what are plainly empirical facts, but there is nothing here which is *guaranteed* to be true no matter what the facts may turn out to be (Putnam 1994f: 248).[9]

More in keeping with the way things are is therefore to say that our empirical knowledge is conventional with respect to certain alternatives and factual with respect to others:

> Saying that there are three objects in the universe of discourse Carnap was describing is a matter of fact, as opposed to saying that there are four objects in that universe, and a matter of convention, as opposed to describing the situation in Leśniewski's language by saying that there are seven objects (Putnam 2004a: 45; cf. also 2012c: 58).

It is therefore because those who follow Carnap and those who follow Leśniewski have made different initial choices that they then answer questions such as the one above in different ways, offering two incompatible ontologies. Moreover, not only the answers concerning the number of objects, but also those concerning the properties of objects depend on the conceptual scheme adopted: if in the world *à la* Carnap at least one of the three objects is completely red and at least one completely black, the statement "There is at least one object that is partly red and partly black" is false for the Carnapian and true for the follower of Leśniewski, according to whom such an object is nothing more than the result of a mereological sum. The property of being partly red and partly black is thus relative to the version of the world we decide to adopt. Moreover: even which entities are "concrete" and which "abstract" can be relative to a version. Consider the dispute over the ontological status of the Euclidean plane that engaged the best scientific and philosophical minds a few centuries ago (Leibniz and Kant among them): are points parts of that plane, as Leibniz held, or mere limits, as Kant claimed in the Second Antinomy of the *Critique of Pure Reason*? That is, are they concrete individuals or abstract entities, say "equivalent classes of convergent series" (e.g., convergent sequences of spheres, as hypothesized above) according to the formalization of geometry elaborated by Alfred North Whitehead and Bertrand Russell in *Principia Mathematica*?[10]

Again, there is no fact of the matter by reference to which the question can be settled. Contrary to Kant's assertion, for Putnam we are not here in the presence of an "antinomy," that is, an impasse of reason determined by the impossibility of resolving a contradiction between two statements, which in turn is determined by the impossibility of grasping things as they are in themselves. As we have already seen, according to Putnam, there are no Kantian and epistemically unattainable "things in themselves," we are not faced with a preordained and pre-constituted world along coordinates unfathomable to the human mind that implicitly establishes which is the best way to describe it. On closer inspection, "the mini-world itself does not *force* us to talk one way or to talk the other way" (Putnam 1994f: 245), it does not fix the ontology once and for all (cf. also Putnam 2012c: 57), it does not predetermine the number of objects and their properties contained within it. And all this applies to the world that human beings concretely deal with: there is no way of describing it that can be considered privileged over others. The cookie cutter metaphor, and with it the reaction of philosophers such as Searle, is wrong because it assumes that in the course of our cognitive enterprise we can clearly distinguish what depends on facts and what depends on our conventional decisions. But, if this were the case, then we would have to be able to answer the question "What are the various parts of the dough?" (Putnam 1987a: 33), that is, the objects considered as completely independent of our representations. Which it is impossible to do: of the mini-world above, for example, we have but two representations, and if we try to answer that the world is made up of three objects, we are merely adopting one of the two conventions (and likewise if we answer that it contains seven). The interpenetration of fact and convention cannot be escaped.

The Nature of the Relation between Equivalent Descriptions

A critic of equivalent descriptions might still retort that this is an unacceptable idea because they bring into play truths that are inconsistent with each other, thus defying logical common sense, and are inconsistent with each other because they are the result of two different conceptual schemes, that is, two different sets of assumptions, theses, and concepts. The diversity of the conceptual schemata within which the two descriptions are embedded therefore prohibits joining them in any relation, much less a relation of equivalence. What's more, if we look closely, the two descriptions are *incommensurable*.

If this is the case, then according to this critic—considered by Putnam close to Quine (cf. Quine 1981: 21–22) and Davidson—the correct thing to say would be that only *one* of two or more descriptions is the true one, while the others are false or, at most, to be considered as mere *façons de parler*—that is, a (supposed) description of things that can only be considered as a "way of speaking" without a genuine connection to reality. This would be the only way to solve the problem of inconsistency, since it would then only be apparent; but—according to this critic whom we could call Davidsonian—it is a solution that only saves equivalence in words, given that this in fact does not exist.

However, even this reaction fails to capture what the phenomenon of equivalent descriptions actually is. To see why, let us finally see what for Putnam is the nature of the relationship that binds two or more equivalent descriptions.

Let us assume that the equivalent descriptions we are considering are two theories. Well, their equivalence is explained by the fact that the primitive terms of the first theory can be defined by means of the primitive terms of the second, so that any statement expressed in the language of the first can be transformed into a statement expressed in the language of the second, and vice versa, without it ever being possible to say that one theory is more fundamental than the other: they have the same explanatory and predictive power with regard to the same portion of the world—that is, they are *cognitively equivalent*.

What assures us of this possibility, Putnam argues, is precisely logic, which allows us to attempt a partially formal definition of the notion of equivalence in question. A very important idea in modern logic is that of the "relative interpretation" of one theory in another: if, for example, one wants to show that a theory is undecidable, then a suitable method is to show that a theory already known to be undecidable is relatively interpretable in the theory in question. It thus follows that it too is undecidable. The definition of *relative interpretation* is given by Putnam in the following terms: a theory T_1 is relatively interpretable in a theory T_2 if

> there exist [. . .] formally possible definitions [. . .] of the terms of T_1 in the language of T_2 with the property that, if we "translate" the sentences of T_1 into the language of T_2 by means of those definitions, then all theorems of T_1 become theorems of T_2 (Putnam 1978b: 38).

The first step in arriving at the desired notion of equivalence then consists in considering two theories to be relatively interpretable with each other: in this way, the notion of *mutual relative interpretability* is obtained. However, this still does not indicate that two theories are equivalent descriptions. Mutual relative interpretability, in fact, is a purely formal relationship between theories, and does not involve the meanings of the terms that appear in them: it only indicates that two theories have a similar formal structure, not that they speak of the same objects, so much so that it is plausible that

two theories about wholly disparate subject matters—say, an axiomatic system of genetics and an axiomatic system of number theory—could turn out to be mutually relatively interpretable, but they would hardly be equivalent in cognitive meaning (Putnam 1978b: 38–39).

Putnam then proposes to combine the formal notion of mutual relative interpretability with the informal notion of *explanation*: in doing so, he ends up defining two equivalent descriptions as two relatively interpretable theories that explain the same phenomena, provided that the "translation" of one into the other preserves the explanation relation. Indeed, "it makes no difference to our predictions or actions which of the two schemes we use" (Putnam 1992b: 117): the theories and sentences of each scheme are perfectly compatible, are "intertranslatable," and thus have the same physical content. Quite simply, "very *different* sentences can describe the very same state of affairs" (Putnam 1992b: 117), ensuring that "an explanation of a phenomenon goes over into another perfectly good explanation of the same phenomenon under these translations" (Putnam 2012c: 57). And to the questions "who *is* to say what is a phenomenon?" about, say, a subatomic particle, and "who is to say what is a perfectly good explanation?" the correct answer is "*physicists* are; not linguists, and not philosophers" (Putnam 2012c: 57).

We are now in a position to appreciate why the Davidsonian objection does not work. First of all, the incompatibility between the equivalent descriptions Putnam speaks of is only so if they are taken "literally" (at face value), only if we pay attention to the "superficial grammar" with which they are formulated, given that it is the mutual relative interpretation that shows that they are *not* genuinely incompatible. And this shows, secondly, that this type of interpretation is not to be equated with the usual cases of translation, those in which two sentences are placed in a relation of synonymy. Statements expressing two equivalent descriptions do not have the same meaning:

> It would be absurd to claim that the *sentence* "there is an electron-wave with the wavelength λ" is *synonymous* with the *sentence* "there is a particle electron with the momentum h/λ and a totally indeterminate position" (Putnam 1967a: 46).

Precisely because the equivalence in question here has nothing to do with synonymy, the notion of meaning is of no use in understanding it. Stating that a certain sentence is relatively interpretable in another

> is not to make a remark about the "meaning" of the words [of these sentences] in any sense of the word "meaning" that an ordinary speaker (or a linguist) would recognize. [. . .] The question "Which translation scheme, if any, preserves the *meanings* of the sentences being translated?" is a bad question. The ordinary notion of "meaning" was simply not invented for this kind of case (Putnam 1994f: 246),

and therefore a question such as the one just seen is nothing more than an attempt "to force the ordinary-language notion of meaning to do a job for which it was never designed" (Putnam 1992b: 118). The Davidsonian objection therefore falls, since it is only possible to speak of inconsistency between two sentences if the question of their meanings and the identity of the meanings (i.e. synonymy) were relevant, and, in the case of equivalent descriptions, this does not hold: "the original sentence and its relative interpretation are *equivalent* without being synonymous, and the two associated conceptual schemes are also equivalent" (Case 1997: 12–13).

But in order to fully understand the epistemological-metaphysical significance of the doctrine of equivalent descriptions and, more generally, of the phenomenon of conceptual relativity of which it is a corollary, it is necessary to appreciate the distinction Putnam posits between, "on the one hand, sameness and difference of *meaning* and, on the other, sameness and difference of *use*" (Case 1997: 13).

Conceptual Relativity

The heart of the conceptual relativity argued by Putnam is that the world that is the object of our descriptions does not impose an univocal use of the expressions of our language—in particular the metaphysically important words such as "object," "individual," "substance," "fact," "state of affairs," "exists," and the like—a use derived from the meaning of such words being fixed in advance and unchangeable. If this were the case, then between, say, the Carnapian's answer and the Polish logician's answer to the question "How many objects exist?" there would be an ineradicable contradiction. On the contrary, the equivalence between the two answers lies in the fact that Carnapian and Polish logician subject terms such as "exists" and "object" to different uses. What occurs in such cases is nothing more than an "extension" of the uses of the words we have called "metaphysically important." It is precisely this that prevents the occurrence of the above-mentioned contradictions and incompatibilities: each statement is in fact placed within a conceptual framework that governs the use of the aforementioned words in its own way, so that *the logical primitives themselves, and in particular the notions of object and existence, have a multitude of different uses rather than one absolute "meaning"* (Putnam 1987b: 97). What's more, "whether such a change of use is or is not a change of 'meaning' is not a question that need have an answer" (Putnam, 1992b: 120). Indeed,

> the question as to which of these ways of using "exists" (and "individual," "object," etc.) is *right* is one that the meanings of the words in the natural language [. . .] simply leave open (Putnam 2004a: 43).

Rather, the use of these words is extended as soon as we adopt a conceptual scheme that proves useful in describing a certain situation, and those who persist in not seeing this

> tr[y] to preserve [. . .] the naive idea that at least one Category—the ancient category of Object or Substance—has an absolute interpretation (Putnam, 1987a: 36).

A correction Putnam makes of an error he had made in the past helps to understand what conceptual relativity is—an error pointed out to him by Jennifer Case (cf. Case 1997: 11). In Putnam 1988 he had given as an example of such relativity the fact that it is possible to describe a room using the vocabulary of elementary particle physics or the vocabulary of tables, chairs, lamps, etc., since—in accordance with his general anti-reductionist stance—he considers them descriptions of equal epistemological and metaphysical dignity. However, closer consideration leads to the conclusion that they are not equivalent descriptions, and thus do not constitute an example of conceptual relativity. Indeed, the two descriptions are neither incompatible, not even literally, nor equivalent from a cognitive point of view:

> the statements "the room may be partly described by saying there is a chair in front of a desk" and "the room may be partly described as consisting of fields and particles" don't even *sound* "incompatible." And they are not cognitively equivalent (even if we do not bar the fantastic possibility of defining terms like "desk" and "table" in the language of fundamental physics, the field-particle description contains a great deal of information that is not translatable into the language of desks and chairs) (Putnam 2004a: 48).

Rather than being an example of conceptual relativity, such cases illustrate what Putnam calls *conceptual pluralism*: the idea that it is permissible to use a plurality of conceptual schemes to describe reality without reducing them to some single fundamental and universal ontology, since—and this will be the leitmotif of Putnam's later work—"the world has many levels of form" (Putnam 2012c: 65).

One last remark before leaving equivalent descriptions and conceptual relativity. Again following Jennifer Case's stimulus, Putnam emphasized a typical aspect of equivalent descriptions, namely the fact that nothing obliges one to make a certain description one's own, to subscribe to a certain conceptual framework and, consequently, to adopt certain linguistic usages: one has the greatest freedom of choice here. This offers the possibility of highlighting how in the language we speak there is an important distinction: that between parts of the language that are obligatory to employ in order to show full mastery of it and the culture that enlivens it, and parts that are merely "optional":

> We are not, given the material and social worlds in which we live, genuinely free not to quantify over tables and chairs, for example. But we are free to employ the conceptual scheme of mereology or not, even in mathematics, even in empirical description, even, for that matter, in philosophy. If we are geometers, we are completely free to either take points as primitive objects [. . .] or to identify them with "limits," and if we do the latter, we are free to take those limits to be equivalence classes of convergent sequences as well! (Putnam 2001: 434–435)

It is optional, and thus a matter of our own free choice, to adopt either the vocabulary proposed by, for example, Leśniewski or that of our standard way of counting. Yet any conceptual scheme that one chooses to adopt has the same validity as any other (assuming that its adoption serves genuine epistemological purposes and that it actually succeeds in explaining the phenomena of the field of enquiry it addresses), making it possible to speak of theories with incompatible ontologies but both correct. But saying this

> is not saying that there are fields "out there" as entities with extension and (in addition) fields in the sense of logical constructions. It is not saying that there are both absolute spacetime points and points which are mere limits. It is saying that various representations, various languages, various theories, are equally good in certain contexts (Putnam 1990a: 41).

Consequently, to ask whether mereological sums really exist, or whether points on the Euclidean plane (or space-time points as a physicist would say today) really exist,[11] or whether an electron is a wave or a particle, is not only meaningless, but downright stupid: God Himself, if He agreed to answer such a question, would say "I don't know!"; and "not because His omniscience is limited, but because there is a limit to how far questions make sense" (Putnam 1987a: 19; cf. also Putnam 1987b: 97).

Above all, by virtue of the interpenetration of fact and convention, in none of the cases just seen is it possible to identify precisely the role played by conventional choices and the role played by facts: there can be no "neutral description," one that reports facts aseptically and with no human contribution, since every description is inevitably "partisan" (Putnam 1987a: 19); but at the same time, such "partisanship" does not imply that the human cognitive enterprise boils down to a mere game of conceptual schemes and optional languages:

> Accepting the ubiquity of conceptual relativity does not require us to deny that truth genuinely depends on the behavior of things distant from the speaker, but the nature of the dependence changes as the kinds of language games we invent change (Putnam 1994g: 309).

As Putnam asks rhetorically:

> why should the fact that reality cannot be described independent of our descriptions lead us to suppose that there are only the descriptions? After all, according to our descriptions themselves, the word "quark" is one thing and a quark is quite a different thing (Putnam 1992b: 122).

Far from taking an idealistic position and making facts a projection of the human mind, Putnam states that "some facts are there to be discovered and

not legislated by us. But this is something to be said when one has adopted
a way of speaking, a language, a 'conceptual scheme.' To talk of 'facts'
without specifying the [optional] language to be used is to talk of nothing"
(Putnam 1988: 114). This is the core of the realist position that we shall see
in Chapters 6 and 7, a position according to which

> to be a *realist who recognizes conceptual relativity* [and the existence of equiva-
> lent descriptions] is to believe that there is an aspect of reality that is indepen-
> dent of what we think at the moment [. . .], and that is *correctly describable
> either way* (Putnam 2012c: 63).

MATHEMATICS AS MODAL LOGIC

As will be recalled, we dealt with Putnam's doctrine of conceptual relativity because I had mentioned that Putnam considers the widespread interpretation of mathematics according to which mathematics is part of set
theory—an interpretation considered a form of *Platonism* due to the fact that
it involves those particular abstract entities that are sets—and his proposal,
modalism, as equivalent descriptions of mathematical reality. Before seeing
in what sense they are equivalent descriptions, and pointing out an important
clarification made a few years before his death, let us finally clarify what
modalism is.

It is an interpretation for which the modal notions of *possibility* and *necessity* are central, in the sense that it is in terms of these notions that mathematical reality is studied. Indeed, Putnam realizes that speaking of mathematical
objects, of sets for example, does not necessarily force one to consider them
as actual (abstract) objects endowed with unconditioned existence, that is,
not dependent on other objects and time, and thus existing autonomously
and eternally in a dimension that is neither physical nor mental, as Platonism
would have it. To begin with, a set vanishes if the elements of which it is
composed vanish, and thus its existence is conditioned by the existence of the
objects (which can also be material) contained in it. But, above all,

> the "objects" of pure mathematics [. . .] are, in a sense, merely abstract pos-
> sibilities. Studying how mathematical objects behave might better be described
> as studying what structures are abstractly possible and what structures are not
> abstractly possible (Putnam 1975f: 60).

What Putnam realized, therefore, is that in mathematics one can say the
same things, that is, one can express the same contents, either by using expressions that refer to (supposedly) abstract objects, or by using expressions that

do not presuppose the existence of the latter at all and that are expressed in the mode of possibility or in the mode of necessity:

> we can avoid quantifying over abstract entities in mathematics entirely, by formalizing mathematics in a *modal logical language*, one which takes as primitive (mathematical) *possibility* and *necessity* (Putnam 2004a: 82).

Geoffrey Hellman, who later developed Putnam's project (cf. Hellman 1989), identified the core of the project in the key idea

> to construe a sentence *S* of ordinary mathematics (concerning, for example, integers or real or complex numbers and functions thereof, etc.) as elliptical for expressing "what would necessarily hold in any model of the appropriate theory involved," where this would take the form of a conditional whose antecedent would contain relevant axioms of the ordinary mathematical theory, with the relation and operation constants replaced with schematic variables, and whose consequent would be the original sentence in question similarly modified (Hellman 2015: 260).

And indeed, the example provided by Putnam consists of starting with the statement, let us call it K, "there exist numbers x, y, z, n (with n greater than 2 and x, y, z different from zero) such that $x^n + y^n = z^n$" and providing a statement that has the same mathematical content but does not speak of the *existence of numbers*, as K does, which immediately may give rise to the description of mathematics as a discipline that describes eternal objects.

Now, since K expresses the negation of the so-called "Fermat's last theorem," Putnam abbreviates the formula expressing K in first-order arithmetic by means of the expression '~ *Fermat*' (where the symbol '~' indicates negation as usual). If ~ *Fermat* is provable, then it is provable from a finite subset of the axioms of first-order arithmetic. Abbreviating therefore the conjunction of these axioms with AX, we have that Fermat's last theorem is false in case the statement $AX \supset$ ~ *Fermat* is valid (where the symbol $\supset$ indicates logical derivation), that is in case this statement is necessary: in symbols, "$(AX \supset$ ~ *Fermat*)." Let us call the latter statement '(1)'.

Putnam points out that if the latter statement is true, its truth does not depend on the meaning of the arithmetic primitive terms. Therefore,

> let us suppose these to be replaced by "dummy letters" (predicate letters). To fix our ideas imagine that the primitives in terms of which AX and ~ *Fermat* are written are the two three-term relations "x is the sum of y and z" and "x is the product of y and z" (exponentiation is known to be first-order-definable from these, and so, of course, are *zero* and *successor*). Let $AX(S, T)$ and ~ FERMAT(S, T) be like AX and ~ *Fermat* except for containing the "dummy" triadic predicate letters S, T, where AX and ~ *Fermat* contain the constant predicates "x is

the sum of y and z" and "x is the product of y and z." Then (1) is essentially a truth of pure modal logic (if it is true), since the constant predicates occur "inessentially"; and this can be brought out by replacing (1) by the abstract schema: [AX(S, T) ⊃ ~ FERMAT(S, T)]—and this [let us call it (2)] is a schema of pure first-order modal logic (Putnam 1967a: 47–48).

We have thus obtained a sentential scheme that simply indicates an entailment relation and does not give the impression of being about objects at all. Not only: (2) is equivalent to K, since they both express the same mathematical content, but certainly the representations they bring to mind are quite different. K induces a picture of mathematics as a description of eternal objects, whereas

> (2) simply says that AX(S, T) entails ~ FERMAT(S, T), no matter how one may interpret the predicate letters S and T, and this scarcely seems to be about "objects" at all (Putnam 1967a: 48).

Putnam's reasoning on the duality "objects-modality," which we have seen concerns a single mathematical statement, can be extended to mathematics in its entirety:[12] it is sufficient to arrive at an analysis of the notion of "standard model for set theory" by making a more conspicuous and expanded use of modal notions. Putnam does not concretely show how to do this, but, as we have mentioned, this is a task that was later accomplished by Geoffrey Hellman. What Putnam does point out, however, is that

> if one fastens on the second picture (the "modal" picture), then mathematics has *no* special objects of its own, but simply tells us what follows from what. If "Platonism" has appeared to be *the* issue in the philosophy of mathematics of recent years, I suggest that it is because we have been too much in the grip of the first picture (Putnam 1967a: 48).

Putnam's move is both simple and revolutionary at the same time: it consists in drawing attention to a cultural-historical, and therefore entirely contingent, fact: that is to say, because of the work of influential philosophers and mathematicians such as Frege, Russell, Zermelo, and Bourbaki, mathematics has been considered quite naturally as concerning a realm of mathematical objects. However, what seems obvious is not always so, and faced with the insurmountable problem of showing how it is possible, not only to refer to abstract objects, but above all how objects without the slightest causal impact on human beings can exist, Putnam shows a completely different way of looking at mathematics—the one according to which there are no typical mathematical objects at all:

You can prove theorems about anything you want—rainy days, or marks on paper, or graphs, or lines, or spheres—but the mathematician, on this view, makes no existence assertions at all. What he asserts is that certain things are *possible* and certain things are *impossible*—in a strong and uniquely mathematical sense of "possible" and "impossible." In short, mathematics is essentially *modal* rather than existential, on this view, [. . .] "mathematics as modal logic" (Putnam 1975f: 70).

Rather than stating, for example, "there exists a set of integers satisfying a certain arithmetical condition," it is more convenient to state "it is possible to choose integers so as to satisfy that condition." Far from being entities existing in themselves, sets are nothing more than "permanent possibilities of selection" (Putnam 1967a: 49; cf. 1975f: 71), as Putnam argues paraphrasing a well-known saying of John Stuart Mill,[13] where this possibility is not a logical, physical, empirical, or technical, but a mathematical one (cf. Putnam 2004a: 67).[14]

Now, on the basis of our analysis of conceptual relativity, we can understand why Platonism and Modalism are for Putnam equivalent descriptions of mathematical reality: on the basis of them, utterances can be formulated that convey the same mathematical content and are intertranslatable. A positive aspect of the existence of equivalent descriptions in cases such as the present is that they represent "a healthy antidote to foundationalism" (Putnam 1967a: 57). This can certainly be considered a good thing, but—as we noted before our *detour* into conceptual relativity—it is surprising that Putnam admits Platonism as an admissible research project, given the weaknesses that it stands accused of. One explanation for this could be that Platonism is after all a "realist" interpretation of mathematics, and that it is therefore precisely for this reason that Putnam keeps it within the ranks of admissible interpretations. This can be indirectly shown by his conviction

that *the internal success and coherence of mathematics* are evidence that it is true under some interpretation, and that its *indispensability for physics* is evidence that it is true under a *realist* interpretation [. . .], in the sense of rejecting all attempts to account for the meaning and the objectivity of mathematical claims in a verificationist way (Putnam 2012g: 219; cf. also 1975f: 74; 2012e: 182),

as we saw at the beginning of the chapter. But there is more than one way to be a realist, and the Platonism/modalism comparison is to the detriment of the former. There are in fact two advantages offered by modalism: one mathematical and the other philosophical. The mathematical advantage lies in the fact that "construing set talk, etc., as talk about possible or impossible structures puts problems in a different focus; in particular, different axioms are evident" (Putnam 1975f: 72), while the philosophical advantage is in turn twofold—both epistemological and metaphysical. The epistemological one we saw at the beginning of the chapter: the aprioristic epistemology that accompanies Platonism can be swept away by the recognition that mathematical knowledge resembles empirical knowledge, and is therefore revisable and not absolute, a recognition favored precisely by modalism. The metaphysical advantage lies in the fact that "it is possible to be a realist with respect to mathematical discourse without committing oneself to the existence of 'mathematical objects.' The question of realism, as Kreisel long ago put it, is the question of the objectivity of mathematics and not the question of the existence of mathematical objects" (Putnam 1975f: 70).

Put it another way, the metaphysical advantage offered by modalism is that

> it yields a natural resolution of Benacerraf's famous problem about "what numbers could not be" and its generalizations. Benacerraf's problem is that although the natural numbers can, as is well known, be identified with sets [. . .], they can be identified with sets in *infinitely many ways*. [. . .] And to stamp one's feet and insist that "the natural numbers are not identical with sets at all" seems a bit arbitrary [. . .] If, however, quantification over sets, functions, and so on is simply quantification over possibilia and not over actually existing entities, then the problem disappears in the sense that all the different "translations" of number theory into set theory, all the different translations of set theory into function theory, and all the different translations of function theory into set theory are just different ways of showing what structures have to *possibly exist* in order for our mathematical assertions to be true (Putnam 2012g: 226).[15]

Thus, although, according to Putnam, Platonism and Modalism are equivalent descriptions, and although the problem of which of the two is true and which false does not impose itself, "the theory of mathematics as the study of special *objects* has a certain implausibility which, in my view, the theory of mathematics as the study of ordinary objects with the aid of a special concept does not" (Putnam 1975f: 72). In conclusion, the fundamental aim of modalism is thus achieved: that "to show a way to be fully realistic [. . .] *but not objectualist* in one's interpretation of classical mathematics" (Putnam 2015e: 279).[16]

EQUIVALENT DESCRIPTIONS?

A volume in the famous Library of Living Philosophers entirely dedicated to the many aspects of his philosophy, and published a year before his death, was an occasion for Putnam to retrace the steps of his career and to clarify certain positions in response to the comments and criticisms he received from a large number of colleagues. Among the various issues he returned to there was the relationship between Platonism and modalism in the philosophy of mathematics, where we can find a useful observation for the purposes of this chapter.

After summarizing the gist of his own definition of equivalent descriptions, after pointing out that most of the examples of this phenomenon he has presented over the years concern the natural sciences, particularly physics, and after pointing out that the translation of one description into the other does not preserve ontology but rather macro-observables and explanations, with physicists, and not philosophers or linguists to be the only ones entitled to say what counts as a good explanation, Putnam notes that, in the case of mathematics, it would be *mathematicians* and not physicists who would have to say what counts as a good mathematical explanation and whether the latter is preserved in the translational transition from one equivalent description to another—for example, in the transition from Platonism to modalism, or from a proof to its formalization in modal-logical language. Here, however, Putnam now realizes, problems arise, "problems that make me think I should never have tried to 'export' the notion of 'equivalent descriptions' from empirical science to the present case at all" (Putnam 2015g: 249).

The reason why is quickly stated:

> Mathematics, after all, is not about "phenomena," but about proofs, ways of conceiving of mathematical problems, mathematical approaches, and much more. And it does not seem reasonable to think that the mapping Hellman and I proposed of mathematical assertions onto modal-logical assertions *preserves* these (Putnam 2015g: 249).

If it is mathematicians who have to say whether the desired equivalence exists between a proof and its formalization in modal-logical terms, then the objection immediately arises that the modal-logical version is not something mathematicians are normally aware of, "and if they were, I doubt very much they would regard them as equivalent, except in the sense of deducible from each other, which is clearly insufficient" (Putnam 2015g: 249). It is thus clear that the *pragmatic* criterion of equivalence proposed in the case of physics cannot be transferred by analogy to the case of mathematics.

There is, however, a way out, and it is this that characterizes the relationship between Platonism and modalism for the late Putnam: the way consisting in considering the "interpretation of arithmetic and set theory as modal statements [in terms of] a *rational reconstruction*" (Putnam, 2015g: 250). What is this all about?

The notion of rational reconstruction has had its fortune in various areas of philosophical reflection, for instance in the philosophy of science. We can cite two of the main exponents of logical positivism, Rudolf Carnap and Hans Reichenbach (I recall that the young Putnam studied with both of them), who set out to offer a reinterpretation of the results of scientists' work that would highlight their implicit metaphysical and epistemological presuppositions, while purging them of potentially misleading aspects. In this way, "intuitive understanding is replaced by discursive reasoning"[17] without regarding "the processes of thinking in their actual occurrence; this task is entirely left to psychology. [. . .] Epistemology thus considers a logical substitute rather than real processes" (Reichenbach 1938: 5). Thus, an attempt at rational reconstruction merely "replaces the rationally opaque psychological processes by which knowledge is typically attained with explicit logical definitions and inferences that show how the results of those processes are genuinely objects of rational knowledge" (Richardson 2006: 682–683): thus, not "a straightforward description [of what scientists say and do], but a way of structuring, of organizing, of rationalizing their actual practice" (Maddy 2007: 78).

As an example of rational reconstruction Putnam does not choose a case explicitly elaborated for this purpose by some philosopher, but invites us to reflect on a question that occupied British mathematicians for almost the entire nineteenth century: that of the reality of complex numbers. None of the various proposals put forward managed to achieve unanimous consensus in the mathematical community, and yet nowadays almost every textbook in analysis tells us

> that we can stipulate that a complex number is simply an ordered pair of members of R, the field of real numbers. And real numbers can be identified with Dedekind cuts on the field Q of rational numbers [. . .]. And rationals themselves can be identified with ordered pairs of members of Z, the ring of the positive and negative integers together with zero [. . .]. And what are the "integers" of which Z is composed? We can stipulate, for example, that they are ordered pairs consisting of a natural number and one of three objects (say, the null set $\emptyset$, its singleton $\{\emptyset\}$, and zero) [. . .]. *And what are the natural numbers?* Well, von Neumann taught us that we can stipulate that they are $\emptyset$, $\{\emptyset\}$, $\{\emptyset,\{\emptyset\}\}$, $\{\emptyset, \{\emptyset,\{\emptyset\}\}\}$, . . . And just as we stipulate definitions for multiplication and addition of members of Z [...], so we stipulate definitions for addition and multiplication of complex numbers [. . .]. Of course, it is necessary to prove that all these stipulations are consistent, and that the distributive, commutative, associative, etc., laws are all forthcoming, but that is straightforward mathematical work. And *voila!* a century of worry by some of the greatest algebraists in the world over the "reality" of, for example, the square root of minus one is *passé* [. . .]. I repeat: an ontological worry about the "existence" of the complex numbers [. . .] is replaced by a mathematical problem—and not that difficult a one—of establishing the consistency and the logical consequences of a set of *stipulations* (Putnam 2015g: 250–251).

Not that the world's greatest algebraists set out to rationally reconstruct their intellectual activity. However, a retrospective look at what they thought and at the results of their theorizing leads one to conclude that some solution to the problem that had engaged them had after all been reached. In other words: rather than a solution, a shift in perspective from which to examine the issue with different eyes. Having started with the task of understanding what kind of "objects" numbers are (and not only complex ones), a certain clarity eventually emerged precisely from the attempt to explain their nature by stipulating a way of reconstructing them logically—even offering a diversity of logical reconstructions of the same objects. It is the simple fact that over time some of the stipulations have become the standard way of referring to those objects—as the manuals of analysis testify—that has made the initial problem "vanish." This is the moral to be drawn from what some algebraists have been doing for decades, if we analyze their activity with the eye of rational reconstruction. The utility of the latter therefore lies in

> "defus[ing]" a metaphysical problem, *not* by showing that there is one right way to think about the issue, but by showing a number of ways we *could have decided to think and talk that would work equally well* (Putnam 2015g: 252).

What, then, is the impact of this on the evaluation of Putnam's modalist thesis? Well, what he proposes with this thesis is nothing more than a rational reconstruction of what mathematicians think and say, that is, "a way of structuring, of organizing, of rationalizing their actual practice" (Maddy 2007: 78), of seeing what lies beneath their statements when they involve certain concepts, in particular the concept of a set. If the aim of a rational reconstruction is to "defuse a paradox" (Putnam 2015g: 253), then the aim is achieved if one succeeds in replacing a problematic concept with a less problematic one: and as we know, modalism points out precisely how "'set' talk can be translated away into *possibility* talk" (Putnam 1978a: 133), showing that "set theory [does] not have to be interpreted Platonistically" (Putnam 2012e: 182). This is one rational reconstruction among (potentially) many, such as Platonism. Both are rational reconstructions, i.e. two ways of speaking, two stipulations, exactly like those of the Carnapian and the mereologist in the classical Putnamian example of equivalent descriptions seen above. If modalism and Platonism now lose the character of equivalent descriptions, they do not lose the characteristic of equivalence. And this reaffirms the idea that Putnam's proposal is not intended to represent the correct epistemological explanation and metaphysical description: "I do not have an 'epistemology' of mathematics (other than to say that one should look at the way mathematicians and physicists actually *do* mathematics)" (Putnam 2015g: 241).

Finally, it is worth noting that the modal conception is what allows Putnam to distance himself once again from Quine. In the literature, their names are in fact paired with reference to the so-called *indispensability argument*, the one that leads from the indispensability of mathematics for science to a realist interpretation of mathematics: from the truth (or the high degree of confirmation) of certain scientific theories would follow the truth (or the high degree of confirmation) of the mathematical theories used by those scientific theories.[18] What is illustrated in this chapter, however, shows how the pairing of the two philosophers is unjustified:

> Quine's indispensability argument was an argument for "reluctant Platonism," which he himself characterized as accepting the existence of "intangible objects" (numbers and sets). [. . .] I was in no way giving an argument for realism about sets [. . .] my "indispensability" argument was an argument for the objectivity of mathematics in a realist sense (Putnam 2012e: 183),

a realism that does not go so far as to postulate the existence of mathematical entities: the existence enjoyed by the latter is a particular kind of logical-mathematical possibility. The mathematician thus studies an objective reality, even if this is not something unconditioned and does not concern immaterial objects; and the physicist who makes use of mathematical statements in formulating a law of nature extracts an actual characteristic of a real material world.

NOTES

1. Cf. Parsons (2015). In Davis (2018), however, Martin Davis argues "that to a greater or lesser extent, all mathematical knowledge is empirical."

2. The idea is that "*something* answers to such mathematical notions as 'set' and 'function'" (Putnam 1975f: 60).

3. Or, rather, in his critique of a *version* of metaphysical realism (the one that characterized the beginning of his career). As we shall see in Chapter 7, there is a sense in which the late Putnam considers himself a metaphysical realist.

4. Mario De Caro has argued, convincingly, that the phenomenon of "conceptual relativity" "could less equivocally be called 'descriptive equivalence,' since the other term may suggest a connection with relativism and antirealism, which would be wholly inappropriate" (De Caro 2022: 257).

5. Before Putnam, it was Reichenbach (cf. 1949: 244) and Quine (cf. 1992: 95 ff.) who drew attention to equivalent descriptions.

6. For other examples of equivalent descriptions, cf. Putnam (1975b): 10ff; (1981): 73–74; (1990a): 39; (1992b): 120ff; (2004a): 66, 81–82.

7. Cf. Putnam (1987a): 33; (1987b): 97.

8. While acknowledging the influence of Carnap (1950) on this point, Putnam emphasizes that, unlike Carnap, he does "not rest the distinction between questions which have to do with the choice of a linguistic framework and empirical questions on an absolute analytic–synthetic distinction. Whether something is or is not 'conventional,' i.e. whether what is at stake is no more than a question of how to talk, is itself something to which empirical facts are relevant" (Putnam 1994a: 247–248).

9. Putnam then hastens to add that where he would criticize Quine is, as we saw in Chapter 1, with regard to the belief that "a distinction between fact and language-choice which is not absolute, not drawn once and for all unrevisably, is of no use."

10. Whitehead and Russell's formalization of Kant's intuitive idea is based on the idea that a "series of spheres is convergent if (1) every sphere (except the first) is contained in the preceding sphere, and (2) the radius of the i-th sphere approaches 0 as i increases without limit," and also on the idea that "two series are *equivalent* if any sphere in either series contains all the spheres after the i-th, for some i, in the other" (Putnam 1992b: 217).

11. It should be noted that "points are not entities we are causally connected with; if one point were removed from space, no physical process, not even the value of the gravitational or any other field at any other point, or the ψ-function at any given point in quantum field theory, would be changed even infinitesimally. All causal explanations are unaffected by the choice between these formalizations" (Putnam 2004a: 46–47).

12. "[. . .] number theoretic statements, with however many quantifiers, can be translated into possibility statements. Thus a statement to the effect that for every number x there exists a number y such that $F(x, y)$, where $F(x, y)$ is a recursive binary relation, can be paraphrased as saying that it is not *possible* to produce a tape with a numeral written on it which is such that if one *were* to produce a Turing machine of a certain description and start it scanning that tape, the machine would never halt" (Putnam 1975f: 71–72).

13. As is well known, John Stuart Mill believed that matter in general, and objects in particular, were nothing more than "permanent possibilities of sensation," an idea, however, with which the realist Putnam can only emphatically disagree.

14. Cf. also: "I am not sure when I learned that 'one can always add one more to any number'—certainly before I was 10 years old. But what sort of a 'can' is this 'can'? I believe that it is the 'can' of *mathematical possibility*. It is not just the 'can' of logical possibility, understood (as it often is) as mere freedom from contradiction; and it certainly had nothing to do with physical possibility. [. . .] 'One can always add one to a number' is not a statement about empirical, or technical, or physical possibility. And yet it is perfectly intelligible, even to most children" (Putnam 2015g: 240–241).

15. However, Putnam hastens to point out that "I do not propose the modal-logical interpretation as a step to arguing that numbers do not really exist, as the title of Hellman's book *Mathematics without Numbers* unfortunately suggests" (Putnam 2012g: 225). As for Benacerraf, cf. Benacerraf (1965).

16. A useful critical discussion of Putnam's modalist proposal can be found in Wagner (2015), Linnebo (2018), Floyd (2020), and Burgess (2018); in the latter also some reservations to the notion of equivalent descriptions.

17. The original text reads: "intuitive Erkenntnis wird durch diskursive Schlüsse ersetzt" (Carnap 1928: §54, 74).

18. For the indispensability argument in Putnam, cf. Putnam (1971): 347. For a discussion of various versions of the argument, cf. Panza & Sereni (2013), Leng (2018). A very useful assessment of the indispensability argument can be found in the six articles (by Concha Martínez-Vidal, José Miguel Sagüillo Fernández-Vega, Otávio Bueno, Sorin Bangu, Susan Vineberg, Matteo Plebani) included in *Theoria* 33 (2), 2018, titled *Updating Indispensability: Putnam in Memoriam.*

Chapter Three

What Is the Human Mind

Let us continue our journey through Putnam's archipelago by examining the way in which the peculiar conception of necessity and apriori that we saw in Chapter 1 reverberates on the analysis of the human mind.

In the philosophy of mind, Putnam is known, on the one hand, for having proposed in the 1960s an interpretation that was destined to be very influential, *functionalism*, so much so that it gave rise to a research program that was carried on by several generations of scholars; and, on the other hand, for having decades later demolished the functionalist hypothesis with at least as much conviction as had motivated its formulation. Let us see how and why such a thing could happen.

THE FUNCTIONALIST HYPOTHESIS

In the first half of the last century, there were two positions that, against Cartesian-style dualism and in accordance with the prevailing general deference to empirical science, occupied a central place in the discussion of the relationship between mind and body: *behaviorism* and the *theory of the identity between mind and brain*.[1] According to the supporters of the former position, only observable behaviors, or at most the dispositions to behavior, constitute the data on which the scientist can reason, and therefore any mental element—internal, private—can rightly be excluded; according to the supporters of the latter, this exclusion is a mistake, since we cannot deny the existence of aspects of our conscious mental life, devoid of the slightest relation to behavior. This recognition, however, identity theorists continue, is not the same as considering the mental aspects of our existence as non-physical elements: far from possessing a spiritual character, they are identifiable in our neurophysiology. Having pain, for example, is nothing other than having a more or less intense stimulation of the C-fibers.

This is in essence the so-called *type identity theory*, which identifies each type of mental state or event with a type of central nervous system state or event.[2] It is a theory that entails a very bold—and therefore potentially very weak—thesis, namely that, given the identification of a certain type of mental state M with a certain type of brain state C, every organism capable of being in M—for example, experiencing pain—can only be in C (and no other type of brain state). This implies the possibility of formulating a psychophysical law that, in accordance with its character as a law, claims to register a regularity without exceptions—the identity of M and C—subsuming under itself a potentially infinite set of instances of M. All this, however, represents a pretty strong (and unjustifiable to boot) constraint on the biology of any organism capable of experiencing pain: indeed, there is nothing to prevent organisms with a very different neurophysiological make-up (e.g., mollusks)—necessarily unable to be in a certain brain state—from also experiencing pain. If, therefore, the model that type-identity theorists had in mind is that of the human brain, we can say that they have committed the sin of "species parochialism."

On the criticism of this thesis, as well as of behaviorism, Putnam began to frame his own position in the philosophy of mind. He considered the focus on manifest behavior and the dispositions to it to be insufficient for a plausible account of our mental life. One only has to imagine a community of *super-spartans* or *super-stoics* that is trained generation after generation to endure even the most acute pain and to suppress all behavioral manifestations of it, to realize the explanatory limits inherent in behaviorism (cf. Putnam 1963: 332 ff.). On the other hand, the sin of parochialism committed by type-identity theorists speaks for itself: the error lies in focusing attention exclusively on human biology—on C-fibers in the case of pain. And the error would consist in exclusively identifying the ability to feel pain with the possession of C-fibers. It would then suffice to focus attention not on the C-fibers as such, but on the role they play in the overall functioning of the organism in which they are found. Here, then, is an avenue that Putnam sees as promising: to pay attention not to what the C-fibers are, but to what they do, to the *function* they fulfill. If function is what matters, then what identifies a mental state M is not its physical substratum, but rather the functional relations that M has with other mental states, with sensory stimuli and with behavioral responses—relations of a causal kind. This is the *functionalist hypothesis* presented by Putnam in the early 1960s.

Putnam's change of perspective was no small one: if an adequate conception of psychology is not about "the what" but "the how," then it is plain that in principle any system, even an artifact, is capable of exhibiting a certain mental state—provided that this system possesses a sufficient degree of functional organization. In short, our psychology is more like the software of a computer than its hardware. Putnam's original equation—rather surprising at the time he introduced it into the debate—between the human mind and a computer program, stemmed from the analogy that can be discerned between psychological states and the logical states of a Turing machine. But what is a *Turing machine*?

TURING MACHINE AND PROBABILISTIC AUTOMATON

Engaged in a program to find a method for deciding whether any mathematical statement is provable or not (a program he succeeded in proving unfeasible), the English logician and mathematician Alan Turing identified computability with "effective calculability," and considered a function to be effectively calculable if its values could be established by a purely "mechanical" process (cf. Turing 1936; De Mol 2018; Hodges 2019). He then described an actual calculation procedure in the form of an ideal machine—which was to go down in history as the *Turing machine*—capable of performing any mathematical operation in a "mechanical" manner. It is an *ideal* calculator because Turing assumes that it does not have the limitations of physically realizable calculators, namely a fixed memory, however large, a finite duration for calculation processes, and inevitable deterioration: a Turing machine, on the contrary, has an infinite memory and perfect, eternal efficiency. What this amounts to is quickly said: it is composed of an infinite tape (in both directions) which is divided into squares (each capable of carrying exactly one symbol), of a scanning head capable of moving from square to square reading their contents, possibly erasing it, writing a certain symbol, or leaving the box unchanged. Each operation of the machine is determined by the state the head is in at a given moment, what the head reads in the box, and the calculation instructions. These instructions—which vary from machine to machine—constitute the *transition table*, which specifies the calculation (an arithmetic addition, for example) that the machine must perform with the symbols it reads in the squares. The basic idea is that, if a certain function is computable, then there is a Turing machine that can compute it in a finite number of steps (as Turing would later show, there are unsolvable problems and numbers that cannot be computed: in such cases, the machine designed to solve those problems never stops and continues to compute indefinitely).

Turing thus emphasizes an aspect that would later be exploited by artificial intelligence theorists and cognitive scientists in general: the fact that an organism's "intelligent" capacities are based on the processing of information, and that this consists of calculations that operate on mental representations. This is an aspect that was initially used by Putnam with some caution: indeed, in his first essay devoted to the functionalist hypothesis he intends to present only an *analogy* between human minds and machines (cf. Putnam 1960b), while later, more decisively, he proposes to *identify* mental states and functional states (cf. Putnam 1967b and 1967c).

However, as you might guess, the "mechanical" aspect (i.e., rigidly conforming to explicit rules) characteristic of a Turing machine makes it difficult to equate the latter with the human mind: while the passage from one state to another of a Turing machine is univocally and automatically determined in virtue of the machine's blind subjection to its own program (the transition table), there is no strict automatism that forces a human mind to enter a certain state rather than another. This is why Putnam appropriately modifies the notion of the Turing machine by introducing the notion of *probabilistic automaton*, which differs from a Turing machine in that it has a finite memory and passes from state to state in a probabilistic rather than deterministic manner; in addition, Putnam imagines that such an automaton has sensory organs capable of perceiving the situation in its environment, and motor organs capable of producing changes in that same environment. Exactly as in the case outlined by Turing, therefore, there will exist as many probabilistic automata as there are (probability) transition tables that can be specified.

Now, any empirically given system (a person, an animal, a computer, and so on) can be the physical realization of a multiplicity of different probabilistic automata, each capable of performing certain operations in accordance with its own transition table. To account for this, Putnam introduces the notion of a *Description* of a system S, stating that it is a true statement according to which S possesses distinct states $S_1, S_2, \ldots, S_n$ connected to each other and to the sensory inputs and motor outputs according to the transition probabilities specified by a given transition table, so that the latter can be considered

the Functional Organization of S relative to that Description, and the state S_i such that S is in state S_i at a given time will be called the Total State of S (at the time) relative to that Description (Putnam 1967b: 434).

This allows the functionalist hypothesis—for example that experiencing pain is a functional state of the organism—to be formulated as follows:

(1) All organisms capable of feeling pain are Probabilistic Automata.
(2) Every organism capable of feeling pain possesses at least one Description of a certain kind (i.e., being capable of feeling pain *is* possessing an appropriate kind of Functional Organization).

(3) No organism capable of feeling pain possesses a decomposition into parts which separately possess Descriptions of the kind referred to in (2).[3]

(4) For every Description of the kind referred to in (2), there exists a subset of the sensory inputs such that an organism with that Description is in pain when and only when some of its sensory inputs are in that subset (Putnam 1967b: 434).

Probabilistic automata thus inherit a fundamental characteristic of Turing machines, namely that of making a complete abstraction from their possible physical realization: they could be made of electronic material, brain tissue, cardboard, or Sardinian cheese, but if they possess a description of the kind given above, then the corresponding mental state can be attributed to them.[4] Retrieving an Aristotelian thesis, Putnam asserts that what matters is *form*, not substance (cf. Aristotle, *De anima*: 412 a6–b11; Putnam 1975d: 302; Nussbaum 2022), and that, indeed, the fact of directing a large part of intellectual effort to the attempt to ascertain what the actual substance of human beings is—that is, whether we are made of pure matter or whether there is something more, some "spiritual" substance—has done nothing but lead to an impasse in the discussion. In a decidedly anti-reductionist outlook, Putnam maintains that the real question to be addressed is "the autonomy of our mental life," and that the fact of possessing a mind endowed with its own typical activity constitutes "a real and autonomous feature of our world" (Putnam 1975d: 291):[5] there is not only a physical reality, but also a mental one. As can be seen, "the functional-state hypothesis is *not* incompatible with dualism" (Putnam 1967b: 436), even if Putnam's recognition of an autonomous mental sphere "needs no mysteries, no ghostly agents, no *élan vital*" (Putnam 1975d: 303), remaining within a strictly naturalistic perspective.

THE SYNTHETIC IDENTITY OF PROPERTIES

In proposing the functionalist hypothesis, Putnam himself admits a certain degree of *vagueness*. The description referred to in clause (2), for example, should be appropriately supplemented by including in the functional organization of any organism certain mathematical functions—such as a "rational preference" function and a "degree of confirmation" function—that are capable of assigning *utilities* to the various situations in which the organism may find itself, determining the organism's behavior in accordance with the theorems of the probability calculus and the rule "Act so as to maximize the estimated utility." In this way, probabilistic automata could be regarded as *rational agents* ("in the sense in which that term is used in inductive logic and economic theory," Putnam 1967c: 409) capable of learning from experience and "criticizing" the way in which the program establishes the succession of logical states.

The subset referred to in clause (4) should also be better specified, for example by clarifying that the automaton's sensory organs include "pain sensors" that are in charge of transmitting a special subset of the sensory inputs, and by establishing that the rational preference function assigns a high degree of disvalue to these inputs.

However, despite the vagueness, Putnam considers the functionalist hypothesis to be much less vague than rival hypotheses, in particular the hypothesis advanced by identity theorists and behaviorists. Let us see why.

First of all, Putnam clarifies the kind of interpretation that must be given to the "is" that appears in a statement such as "Pain is *K*," where *K* can be replaced by a sentence such as "stimulation of C-fibers," or "a certain behavioral disposition," or "a certain functional state." Contrary to the methodology developed in analytic philosophy around the middle of the last century, aimed at the semantic analysis of terms (including the terms used, and sometimes coined, by scientists), the relationship that in the above-mentioned statement links two properties—the property of "feeling pain' and the property of 'having *K*'—does not solely have to do with the meanings of the expression "feeling pain" and the expression that is substituted for *K*. According to Putnam, the "is" that relates the two properties is not the "is" of meaning analysis, the one that establishes a semantic equivalence—a *synonymy*—between the two expressions flanking the "is." On closer inspection, thinking that two properties can only be identical if the corresponding expressions are synonymous is tantamount "to collapse the two notions of 'property' and 'concept' into a single notion" (Putnam 1967b: 430),[6] an operation that is not always legitimate. Indeed, we can consider—albeit roughly—the meaning of a term as the concept of the object to which the term refers: terms with the same meaning thus refer to the same concept; to a good approximation, therefore, we can take the concept of mare and that of female horse as a single concept—given the synonymy of the corresponding terms. Nevertheless, stating, for example, that the property of being water is the property of being H_2O is not the same as stating that the concept of water is the concept of H_2O, as can easily be seen from the difference between the meaning of the term "water" and the meaning of the term "H_2O." Above all, if, on the contrary, one were to hold that the expressions "water" and "H_2O" are synonymous, this would be tantamount to holding the statement "Water is H_2O" to be *analytic*—that is, to hold it to be true by virtue of the meanings of the expressions composing it, and thus knowable *a priori*. But all this would be a gross error: that experience is needed to know the truth of the statement "Water is H_2O" is all too obvious, just as it is obvious that the concept of water and that of H_2O are different concepts.

How then is the 'is' in 'Pain is K' to be interpreted, given that it is not an analytic identity statement?

As is clear from the example just given, "Water is H_2O," we are here dealing with an identity of a *theoretical* kind, in the sense that the statement in question follows directly from one or more theories that guarantee its truth. The "is" is here the "is" of theoretical identification, the one allowed by the best theories we have at our disposal, those provided by psychologists, neurophysiologists, physicists, chemists—as the case may be. It is by virtue of the presence of a background of accepted theories in a given epoch that a statement such as "Pain is K" does not merely establish a "correlation" between the property of feeling pain and the property K, but an identity in its own right. And not an analytic identity, but a *synthetic* one: "no important theory of the nature of mind can either be confirmed or ruled out by an examination of the meanings of mental words," Putnam declares; and indeed "my view is a synthetic hypothesis, not a contention about the meaning of mental words" (Putnam 1975b: xiii–xiv).

The "is" that appears in clause (2) is thus permissible, if interpreted in the right way—that which specifies a synthetic identity between properties.[7] However, this is also the interpretation that can be ascribed to the identity statements of brain-state theorists and behaviorists: why then, according to Putnam, are their proposals vaguer and therefore less plausible?

CRITICISM OF RIVAL PROPOSALS

The main difficulty that the behaviorist proposal runs into is, according to Putnam, an irremediable *circularity*. Explaining what pain is by claiming that it is "a certain behavioral disposition," and then specifying this behavioral disposition as "the disposition of X to behave as if X were in *pain*" (Putnam 1967b: 438), explains nothing: it would be like explaining that today is a fine day by claiming that the sun is shining and the sky is blue, and then specifying this weather situation as "the situation that occurs when it is a fine day." It is precisely the failure to describe the behavioral disposition invariably correlated with pain (and typical of any organism, regardless of species) without using the term "pain" that tinges the behaviorist proposal with vagueness. Moreover, if the behaviorist were to express his hypothesis solely with reference to manifest behavior, and were to regard only what takes place outside an organism as relevant to the attribution of psychological states, then his hypothesis would not only be vague but downright false. For if two animals have their motor nerves permanently anaesthetized, and only one of them also has its pain fibers permanently anaesthetized, then only one of them will feel pain, but both will have the same (actual and potential) behavior: it is therefore false that any organism that feels pain necessarily has a behavior (actual or potential) typical of pain (remember the case of the super-spartans noted above). Similarly, it is false that any organism that has a behavior (actual or potential) typical of pain necessarily feels pain: just imagine an actor with exceptional acting qualities simulating a perfect state of pain. Finally, Putnam concludes,

> even if there *were* some behavior disposition invariantly correlated with pain (species-independently!), and specifiable without using the term "pain," it would still be more plausible to identify being in pain with some state whose presence *explains* this behavior disposition—the brain state or functional state—than with the behavior disposition itself (Putnam 1967b: 439).

Having set aside the behaviorist proposal, however, that of brain-state theorists remains in the field. It is at this point that Putnam invokes the idea of species parochialism mentioned above. Let's stick to the case of the mental state of pain. According to Putnam, the identity theorist must not only be able to specify a brain state in which any organism that feels pain should be and in which any organism that does not feel pain should not be, but must also be "nomologically certain" that it is a brain state of any terrestrial or extraterrestrial entity that can be discovered in the future and that is capable of feeling pain. And all this makes such a hypothesis very *ambitious*. Moreover,

> the hypothesis becomes still more ambitious when we realize that the brain-state theorist is not just saying that *pain* is a brain state; he is, of course, concerned to maintain that *every* psychological state is a brain state. Thus if we can find even one psychological predicate which can clearly be applied to both a mammal and an octopus (say "hungry"), but whose physical-chemical "correlate" is different in the two cases, the brain-state theory has collapsed. It seems to me overwhelmingly probable that we can do this (Putnam 1967b: 436–437).

On the contrary, the Putnam of the 1960s argues, the functionalist hypothesis appears more analyzable than its rivals, both mathematically and empirically. Given that one generally ascribes a certain mental state to an organism by observing its behavior, it is fair to say that the similarities found in the behavior of two systems constitute a good reason to suspect similarities in their functional organization, and—instead—a much weaker reason to suspect similarities in their physical constitution. Besides that, at least as far as the fundamental psychological states (feeling hunger, thirst, anger, etc.) are concerned, there are good reasons to believe that the transition probabilities of these states with each other and with behavioral manifestations are the

same in organisms belonging to different species, since this is how we identify such states; indeed, we would not regard a certain animal as hungry if it
did not have a behavior directed toward food and this was not followed by
a state of satiety. In contrast to the brain-state theorist, who has to hope for
a future formulation of species-independent neuro-physiological laws, the
functionalist has only to hope for a future formulation of species-independent
psychological laws, and this—in addition to emphasizing the aforementioned
autonomy of mental life and, hence, of psychology—makes the functionalist
hypothesis not only less vague and more analyzable but also more explanatory than its rivals:

> Understanding why the machine, say, computes the decimal expansion of π,
> may require reference to the abstract or functional properties of the machine, to
> the machine's program and not to its physical and chemical make up (Putnam
> 1975b: xiii).

And this, Putnam concludes, is the way along which to fulfill what seems
to be the task of psychology: "to produce 'mechanical' models of organisms"
(Putnam 1967b: 435).

THE DEPARTURE FROM COMPUTATIONAL
FUNCTIONALISM: LIBERALIZED FUNCTIONALISM

As early as the early 1970s, Putnam became aware that the analogy between
the human mind and a Turing machine did not hold. Indeed, as we have seen,
a Turing machine can be in one state at a time, and these states are "instantaneous," that is, they represent the total situation of the system at a given
time and dictate the next state. Moreover, learning and memory do not, by
themselves, determine new states of the machine: they are not represented as
the acquisition of new states, but rather as the acquisition of new information printed on the tape. The behavior of the human mind, however, is quite
different: firstly, learning and memory play an important role in determining
its states; secondly, these are not instantaneous, they do not alone and univocally fix the next mental state. Above all, a mental state does not represent
the total situation of the mind at a given time, since multiple states do so
simultaneously.

All of this highlights what Putnam sees as a negative aspect of machines
in the philosophy of mind: while they undoubtedly play a positive role in that
they induce an appreciation of the difference between abstract structure and
its many concrete realizations, stressing how their most important properties
are not physico-chemical, they also lead to an unwarranted oversimplification,

implicitly ventilating the idea that human functional organization can be as narrow and well-determined as that of a machine.

Nevertheless, bear in mind that the critique of Turing machines as a model of the human mind is not yet equivalent to a critique of functionalism: in fact, "functionalism is committed to the claim that mental states are computational states, not to the Turing machine model" (Shagrir 2005: 230). However, a crack had opened in Putnam's philosophical support for the functionalist hypothesis. One of the virtues of the latter, as he emphasized from the outset, was, as will be recalled, the anti-reductionist spirit, condensed in the rejection of any reduction of the mental sphere to any other sphere to be considered more basic—to a physico-chemical substratum, for example. But Putnam soon realized that even functionalism was not without a reductionist streak: after all, the identification of mental states with computational states was nothing more than a reduction of the former to the latter, a reduction that psychology was initially intended to bring about as soon as it had completed its program—"to produce 'mechanical' models of organisms," as we have seen. Here is a passage from Putnam's own reconstruction of this decades later:

> Once psychology has progressed far enough in the pursuit of this "inevitable program" to make the hypothesis that mental states are just functional states precise, it will be possible—or so I claimed—to confirm the hypothesis in a way analogous to the way in which we have confirmed theoretical identifications in physics (Putnam 1994c: 509).

However, both in the idea that psychology has the task of producing a mechanical model of any organism (not only human and not only natural), and in the related idea that it is possible to confirm psychological hypotheses in the same way as physical hypotheses are confirmed, there is an *utopian* vein, which in turn derives from a *scientistic* attitude—an attitude of excessive and, in the end, uncritical deference to science and its potentialities. *Naturalism* in a strong form. Part of the maturation and changes through which Putnam's philosophy passed was due to the detachment from the belief that a certain and definitive resolution of problems, including philosophical ones, necessarily passes through an application of the scientific method (cf. Putnam 1992g: 350). To think that science is capable of solving all problems, and in particular that psychology can be made "scientific" by virtue of its simple combination with, say, computer science, is nothing but utopia in its pure state.

Anti-reductionism, anti-scientism, anti-utopianism, and a finer naturalism—all threads in the same complex skein—are certainly factors that cooperated to make Putnam abandon functionalism. But the decisive blow to functionalism came from one of Putnam's most important achievements in the philosophy of language, namely that the meaning of most of our words, as well as the content of our thoughts, are *not* determined simply by our functional organization—that is, they do not depend solely on what goes on inside our heads. The content of our mental life depends mostly on the world outside our minds, and it is this belief that—although already clear to Putnam in the early 1970s—had full influence in determining the abandonment of computational functionalism some thirty years later (cf. Ben-Menahem 2022: 293). According to Putnam, in fact,

> It is in the context of a network of social and physical interactions, and only in such a context, that I can do such a thing as "think that the price of gold has become very high in recent years." If thinking that thought is what I once called a "functional state," it is not (as I mistakenly believed) simply a "computational state" of my brain; the "function" in question is a world-involving function (Putnam 2016a: 210).

Paraphrasing a slogan he made famous in the 1970s, he summed up his point by stating that "the mind isn't in the head" (Putnam 2016a: 210).[8]

The result is a kind of liberalization, that is, an "(environment-involving) 'liberal functionalism'" (Putnam 2012d: 637):

> For a liberalized functionalist, there is no difficulty in conceiving of ourselves as organisms whose functions are, as Dewey might have put it, "transactional," that is, environment involving, from the start (Putnam 2012b: 83).

The conviction that we are dealing here with "transactions" is important, because it highlights how identifying *sic et simpliciter* mental states with computational states is nothing but science fiction, and emphasizes how mental states are the result of a complex interaction between us and the environment, and thus depend "on our nature as well as the nature of the environment" (Putnam 2012d: 636). In Chapter 7 we will take up this conception of the mental faculties as having "long arms" (Putnam 2012b: 83), and of the mind as a *system of abilities* and not as an organ in itself. Here, in conclusion, we draw attention to the fact that a mind understood in this way—a mind whose operative and interactive abilities are nothing but multifaceted ways of functioning in continuous contact with the surrounding environment and realizable in constitutionally different systems—is not exhausted by internal capacities of computation, does not admit reductivist descriptions based solely on the vocabulary of some science (e.g., computer science), but broadens this vocabulary to include the vocabulary of psychology, biology, neurology, as well as the normative and intentional vocabulary of semantics.

Simply put, "semantic externalism implies externalism about the mind" (Putnam 2016a: 223). But what exactly is *externalism*?

NOTES

1. Behaviorism was linked to distinguished names such as Rudolf Carnap, Carl Gustav Hempel and Gilbert Ryle, while identity theory was at the time mainly advocated by Ullin Place, Herbert Feigl, and J.J.C. Smart.

2. Another possibility for proponents of the mind-brain identity theory is to interpret the two equalized entities not as general "types," but as particular "occurrences" or "tokens" of a type: a specific token of a mental state occurring at a certain place and time, on the one hand, and a specific token of a brain state occurring at the same place and time, on the other. This is a less strong version of the theory, since the identification of a single specific mental event (e.g., the migraine I felt last New Year's Eve) with a single specific neural event (e.g., the stimulation of my C-fibers) does not preclude that the same mental event happening now in the person in front of me can be identified with an entirely different brain event (e.g., the stimulation of her amygdala), making it impossible to formulate psychophysical laws.

3. With this clause Putnam intends to prevent "social organisms," such as swarms of bees, from being considered capable of feeling pain.

4. This is a version of the *multiple realizability thesis of mental states*, typical of any version of functionalism: given a mental state K, it can be realized by different physical substrata (or, even, by nonphysical substrata such as, assuming it exists, a disembodied spirit), provided that K has an appropriate functional description (cf. Putnam 1960b: 371).

5. Cf. also: *"What is our intellectual form?* is the question, not what the matter is. And whatever our substance may be, soul-stuff, or matter or Swiss cheese, it is not going to place any interesting first order restrictions on the answer to this question" (Putnam 1975d: 302).

6. Putnam attributes this "fusion" of concept and property to Carnap (1947).

7. For a more extensive treatment of this topic, cf. Putnam (1969b). Cf.: "What first led me to write a paper on the topic of 'properties' was the desire to study reduction in the case of *psychology*" (313).

8. Evidence of the fact that as early as the 1970s it was clear to Putnam that "the mind isn't in the head" can be gathered from Putnam 1974–75: 298. This idea came to him from John McDowell, and he had to wait some twenty years to become fully aware of its implications. As for McDowell, cf. for instance: "the moral of Putnam's basic thought [that at least some meanings are at least in part environmentally constituted] for the nature of the mental might be, to put it in his terms, that the mind—the locus of our manipulations of meanings—is not in the head either. Meanings are in the mind, but, as the argument establishes, they cannot be in the head; therefore, we ought to conclude, the mind is not in the head" (McDowell 1992: 36). We return to this point in Chapter 7.

Chapter Four

The Causal Conception of Meaning

If you think that the contents of human minds—our beliefs, our intentions, and what we mean by our verbal expressions—depend only, or mostly, on properties of our body, typically the brain, then you are an *internalist*. Instead, if you think that those contents depend on something outside your body, typically the social and natural environments, so that causal interactions with items in both environments play a crucial role in content individuation, you show that you are subscribing to *externalism*.[1] Putnam's philosophical reflection gave an enormous contribution to the spelling out of externalism in epistemology, metaphysics, philosophy of mind, and philosophy of language, so that it is no wonder that when, in 2011, he was awarded the Rolf Schock Prize—the equivalent of the Nobel Prize—in Logic and Philosophy by the Royal Swedish Academy of Sciences, the motivation was "for his contribution to the understanding of semantics for theoretical and 'natural kind' terms, and of the implications of this semantics for philosophy of language, theory of knowledge, philosophy of science and metaphysics."

The Swedish Academics got it right: their motivation could be taken to embody the core of Putnam's entire philosophy, and the contribution they refer to is externalism. Indeed, if—as we have seen—naturalism is the hallmark of Putnam's philosophy, externalism is attractive to Putnam's eyes just thanks to

> its consonance with a naturalistic way of thinking about human beings in the world. The idea of us as beings with an intrinsic ability to think about things whose properties we may neither have experienced nor detected with the aid of scientific instruments nor have the ability to define in terms of properties we have interacted with in one of these ways makes reference a magical power, and I believe that referring is a perfectly natural affair (Putnam 2016a: 222).

In this chapter we address the "semantics for theoretical and 'natural kind' terms"; in Chapter 5 its "implications [for] philosophy of science"; and in the following chapters its implications for metaphysics, the theory of knowledge, and—again—philosophy of mind. So, let's see how, according to Putnam's semantic externalism, *reference depends on causal connection to the extra-bodily environment* (Putnam 2016a: 218).

THE SEMANTIC TRADITION IN A NUTSHELL

In the 1970s Putnam published a series of papers devoted to the analysis of the meaning of terms of physical magnitude and natural kind,[2] culminating in the essay "The Meaning of 'Meaning'"—"probably the most widely read and cited of any he ever wrote" (Floyd 2005: 21)—a sort of *summa* of the results achieved with the previous papers, in which he shows the repercussions that his semantic conception has on a general philosophical level.

The opening paragraphs of "The Meaning of 'Meaning'" represent the inevitable *pars destruens* intended to clear the ground of theses deemed implausible and prepare it for new foundations. Having recalled the custom of distinguishing—in the meaning of a linguistic expression—between *intension* and *extension*, Putnam begins by noting that while the notion of extension is sufficiently precise thanks to the logical notion of truth,[3] that of intension has not received an equally good specification other than the vague notion of "concept." Hence the idea that meaning (in the sense of intension) is a *mental* entity, since in the philosophical tradition concepts have mostly been conceived as something mental.[4] Certainly, eminent philosophers such as Frege and Carnap had reacted vigorously against the inclusion of meanings in the psychological sphere, since they correctly emphasized the fact that meanings cannot but be "public," that is, graspable by several people or by one and the same person at different times, and thus not the exclusive property of this or that private mind; rather, they argued, concepts and meanings are *abstract* entities. However, Putnam notes, grasping an abstract entity is an individual *psychological* act, an act by which "speakers are supposed to be able to direct their mental attention *to* concepts by means of something akin to perception" (Putnam 1988: 129): understanding a term is therefore equivalent to being in a certain psychological state. In spite of Frege's and Carnap's move, then, psychology does not seem to have been eliminated from traditional semantics: from Aristotle[5] to Frege and Carnap, via at least John Stuart Mill and Bertrand Russell, the basic thesis is that "there is something in the mind that picks out the objects in the environment that we talk about. When such a something (call it a 'concept') is associated with a sign, it becomes the meaning of the sign" (Putnam 1988: 19). Accordingly, here is a principle that Putnam believes underlies traditional thinking about meaning: "(I) That knowing the meaning of a term is just a matter of being in a certain psychological state" (Putnam 1975e: 219).[6]

Another assumption, taken for granted in secular thinking about language, is that two terms cannot differ in extension and have the same intension.[7] If the intension is the same, the extensions cannot differ. From here one can easily derive the principle—the second one—"(II) That the meaning of a term (in the sense of 'intension') determines its extension" (Putnam 1975e: 219).[8] The main purpose of the opening part of "The Meaning of 'Meaning'" is to show that there is no notion—let alone a notion of meaning—that can satisfy both principles: "the traditional concept of meaning is a concept which rests on a false theory" (Putnam 1975e: 219), a theory that for simplicity we might call the *Aristotelian perspective*. To achieve this, Putnam criticizes a consequence of the two principles, namely the thesis that the psychological state a speaker is in when using a certain term determines the extension of the term,[9] and he does so by formulating a science fiction example destined to be cited and discussed in various forms to this day: the *Twin Earth example*.[10]

TWIN EARTH

Let us imagine that somewhere tucked away in the Universe there is a planet exactly identical to ours except for one detail: that liquid which we call "water" and which on Earth has the chemical composition H_2O has, on this planet, a different chemical composition, say XYZ. In saying that the planet is exactly identical to ours I mean that for every object, person, characteristic present on Earth there is an analogue on this planet. There is a people identical to the U.S. people, and a doppelganger for each U.S. person, who speaks English, a people identical to the French people, and a doppelganger for each French person, who speaks French, a people identical to the German people, and a doppelganger for each German person, who speaks German: it is a Twin Earth (TE) to all intents and purposes, except for the chemical composition of that liquid that, as on Earth, is used to quench one's thirst, wash oneself, cook spaghetti, that fills lakes, rivers and seas, that English-speaking Twin Earthers call "water" and "the French-speaking ones call [. . .] 'eau,' and the German-speaking ones call [. . .] 'Wasser'" (Putnam 2016a: 208–209). It should be noted that if one were to judge only by external appearance, the two liquids would be completely indistinguishable, so much so that Earthlings who were transported by spacecraft to the TE would not notice any difference and would easily call "water" the liquid they would be offered to drink or use to wash themselves.

Of course, it is likely that if there is a chemist among the Earthlings visiting TE who, for whatever reason, analyzes a sample of that liquid and realizes that it is not H_2O, she will have qualms about behaving with that liquid exactly as she behaves with water on Earth. A natural reaction for her might be to drink it with some apprehension, or to state "On TE water is not water," or "On TE the word 'water' means XYZ," or "On TE there is a liquid that plays the role of water, but it is not water." In short, the mental state she is in now when she behaves (verbally and non-verbally) with that colorless, tasteless, thirst-quenching, useful-for-washing, etc., liquid, is different from the mental state she was in when she behaved (verbally and non-verbally) with water on Earth (even though she uses the same word in both cases).

Suppose now that a time machine transports us to a time when chemistry was not developed to the point where the difference in the chemical composition of the two liquids could be understood—say around 1750. None of the Earthlings who landed on TE would have been able to notice any difference, and therefore the mental state in which an Earthling was in using the word "water" and that of a Twin-Earthling using the same word would have been perfectly identical: all of one's beliefs related to water would have been the beliefs of the other, and vice versa. Yet the word extension of the Earthling would have consisted of all samples of H_2O, just like today, while the word extension of the Twin Earthling would have consisted of all samples of XYZ, just like today. And this goes to show that the extension of a word is not determined by the psychological state that speakers are in when they use that word: the above consequence of the two cardinal principles of traditional semantics is false.[11]

Note that this conclusion does not necessarily depend on putting things in a science fiction context. Putnam invites us to reflect on the fact that many of us happen to be unable to distinguish between an elm and a beech: I, for example, might be able to tell that a tree placed in front of me is either an elm or a beech, but I could not tell for sure which. Everything I can believe about beeches applies to elms and vice versa, which shows that my psychological state when I think of beeches and use the word "beech" is identical to my psychological state when I think of elms and use the word "elm." Yet, in my idiolect "beech" refers to beeches and "elm" to elms, since that is how these words entered my personal lexicon.[12] Again, psychological states do not determine the extensions of terms: in the same way that, having found out the way things are, an Earth chemist would say "On TE the word 'water' means XYZ," so in my idiolect "beech" means beech and "elm" means elm, even though in both cases there is no difference in mental states: "cut the pie any way you like, 'meanings' just ain't in the *head*!" (Putnam 1975e: 227)[13]

BUT THEN WHERE ARE THE MEANINGS?

A first partial answer to this question can be gleaned from the elm/beech example just seen. I may not know how to distinguish an elm from a beech, and thus not use the relevant words correctly in all contexts, but I still have the option of turning to a botanical expert in cases where correct discrimination is of particular importance to me. In the mind of the individual speaker there may not be necessary and sufficient criteria for making such a distinction, but the knowledge of these criteria is still the heritage of the entire linguistic-cultural collectivity to which I belong, and this, according to Putnam, shows that alongside the well-known "social division of labor" there is a *social division of linguistic labor* that allows the average speaker to be unaware of the criteria for fixing the extension of part of the words he uses, and yet allows her to use those words with the correct extension: indeed, what fixes the latter is "the sociolinguistic state of the collective linguistic body" (Putnam 1975e: 229). If you want to identify meanings, it is first of all to the society to which a speaker belongs that you must look:

> Language is a form of cooperative activity, not an essentially individualistic activity. Part of what is wrong with the Aristotelian picture is that it suggests that everything that is necessary for the use of language is stored in each individual mind; but no actual language works that way (Putnam 1988: 25).

As is well known (the obvious reference is to Wittgenstein 1953: §11), words are tools, but (somewhat at odds with Wittgenstein) they are not comparable to hammers or screwdrivers, which can be used by one person in isolation. Rather, words are "like a steamship which require the cooperative activity of a number of persons to use" (Putnam 1975e: 229). I may not be able to distinguish an elm from a beech, or a gold ring from a pinchbeck ring, but to think that this shows that I don't know the meaning of the words "elm" and "gold" amounts to "confus[ing] lack of botanical [or metallurgical] knowledge with lack of linguistic competence" (Putnam 1996: xvi). In brief,

> knowing the meaning of the word "gold" or of the word "elm" is not a matter of *knowing that* at all, but a matter of *knowing how*; and what you have to *know how* is to play your part in an intricate system of social cooperation (Putnam 1996: xvi).[14]

This, however, is only a first part of the answer to the question posed in the title of this section. There is a second and far more fundamental one. Before we see what it consists of, let us try to answer another question. We have said that (in most cases) I may not know what the object (or set of objects) to which a natural kind word refers actually is, and yet—by virtue of the social division of linguistic labor—be able to use that word with the right reference. But how did I acquire this ability to refer to the right objects using that word?

THE CAUSAL CHAIN OF REFERENCE

In the very years that Putnam was maturing his ideas about the meaning of terms of physical magnitudes and natural kinds, logician and philosopher Saul Kripke focused on analyzing proper names along lines that were to reveal more than one point of contact with Putnamian ideas.[15] At least with regard to two aspects Putnam acknowledged a debt to Kripke: one is the way in which a speaker acquires the ability to refer correctly to a certain individual or class of individuals by using a term: this happens through a *causal chain of communication* (or *reference*).[16]

When a name is given to a child, an animal, a hitherto unknown plant, a newly discovered subatomic particle, or a newly manufactured object, a *naming ceremony* occurs, the effect of which is the introduction of a name into language and its attribution to object, particle, plant, child, and so on. The verbal and non-verbal course of action of one or more people causes a previously absent situation. Subsequently, the verbal interactions of "baptizers" with other speakers cause the word used for baptizing to be transmitted and spread through the language community: those who learn the correct usage from the baptizers in turn transmit that usage to other speakers, forming a chain in which each speaker represents a link connected to the next link and to the previous one by a causal relationship. Here is Kripke:

> Through various sorts of talk the name is spread from link to link as if by a chain. A speaker who is on the far end of this chain, who has heard about, say, Richard Feynman, in the market place or elsewhere, may be referring to Richard Feynman even though he can't remember from whom he first heard of Feynman or from whom he ever heard of Feynman. He knows that Feynman is a famous physicist. A certain passage of communication reaching ultimately to the man himself does reach the speaker. He then is referring to Feynman even though he can't identify him uniquely (Kripke 1972: 91).

By virtue of such a chain, a speaker can use a word with the right reference even if she cannot provide a complete and exhaustive description of the referent, just as happens to me with the word "elm." Basically, what is essential for a speaker to be able to use a word correctly are three things: being part of a causal chain, possessing a minimal amount of information about the referent (the minimal amount that runs through the chain like a stream) and "intend[ing] to use it with the same reference as the man from whom he heard it" (Kripke 1972: 96): this last aspect (the *referential intention*) is of paramount importance, on pain of breaking the chain and (possibly) setting up a different chain of communication.

Note that it is precisely because the average speaker is obligatorily required to possess only a minimum amount of information about the referent of a certain term that, in controversial cases, she is forced to turn to experts. Indeed, they are in possession of more detailed and, in the case of terms of physical magnitude or natural kinds, scientifically precise information thanks to which, say, we can understand whether the object we are observing in the laboratory is actually an electron and whether we are using the relative term correctly. As we have said, the meaning of the term "electron" is in the sociolinguistic body and not in the mind of the individual speaker; we have also said, however, that this is only part of the story, and that in reality for Putnam the meanings of the terms he examines lie elsewhere. To understand where, let us reflect on what we have said about TE, keeping in mind that we have yet to see what Putnam's second debt to Kripke is.

TWO ALTERNATIVE HYPOTHESIS ABOUT THE MEANING OF "WATER"

With the example of TE Putnam has shown the falsity of one consequence of the two cardinal principles of traditional semantics—the Aristotelian perspective, as we have succinctly called it. The question now is which of the two principles is to blame for the falsity of the thesis that the psychological state a speaker is in when using the word "water" determines its extension. Essentially, having noted the falsity of this thesis and having ascertained that there is H_2O on Earth and XYZ on TE, two possible explanations of the meaning of the word "water" remain on the table. One could argue either that

(1) the word "water" has a *constant meaning*, and thus that it means on Earth what it means on TE, only that water is H_2O on Earth and XYZ on TE; or that

(2) it has a *relative meaning*, and therefore that it does not mean on Earth what it means on TE: water is H_2O in all possible worlds, and therefore on TE there is no water, even though there is the word.

Which of the two explanations is the right one? The answer reveals a strategy typical of Putnam's way of doing philosophy: that of paying close attention to "facts which are, so to speak, under our noses" (Putnam 1988: 24), of privileging the verbal and non-verbal reactions that we spontaneously display in the most disparate situations of our existence, in the belief that what we normally do can sometimes have unsuspected philosophical importance.

Indeed, what we are considering here is "a topic which deals, after all, with matters which are in everyone's experience, matters concerning which we all have more data than we know what to do with, matters concerning which we have, if we shed preconceptions, pretty clear intuitions" (Putnam 1975e: 271).

Meanwhile, let us emphasize the fact that hypothesis (1) amounts to the negation of the second principle of traditional semantics (that by which intension determines extension) and the preservation of the first (knowing the meaning of a term means being in a certain psychological state). Second, let us consider what we mean implicitly when we indicate a glass of water by saying "This liquid is water," for example in the case where we want to teach someone the use of the word or simply give them information. Our statement implies an empirical assumption, namely, that that sample of liquid in the glass is identical to most of the substance that we and other speakers in our language community have at other times called "water:" the substance present in our natural environment and with which we have been causally connected since birth by virtue of a myriad of direct and indirect interactions—H_2O. When we use the word "water," we cannot but mean the liquid present in our world: everything that counts as water to us, therefore, can only be H_2O, and everything that does not have this chemical composition is not water to us. In short, for us water is H_2O in all possible worlds. No matter how far we go in our galaxy, water would not cease to be H_2O for us: all our referential intentions when we use the word implicitly point to the substance which is chemically identifiable as H_2O, and this indicates the centrality of that substance in determining our use of the word itself—thus in determining its meaning. In short, the correct explanation of the meaning of the word "water" is the second one above.

To reiterate: all this means, in the final analysis, that what determines the meaning of a term of natural kind or a term of physical magnitude is the entity to which the term refers, the entity present in the natural environment with which a linguistic community is causally connected: meanings, already removed by Putnam from the mind of the speaker by tracing them in the social body, are in fact found even further away: *in the world.*[17]

THE MEANING OF "MEANING"

The rejection of explanation (1) brings with it the denial of the first principle of traditional semantics in favor of the second. The latter, however, is preserved in an obvious way, since if the meaning of a natural kind term is given by the set of *things* that constitutes the extension of the term, we can say that, inevitably, the meaning determines the extension.[18] Thus, Putnam finds himself in a position to propose a description of the meaning of any term of natural kind or physical magnitude. Such a description consists of a *vector*, that is, a finite sequence whose components are at least four (Putnam leaves open the possibility that further analysis will identify others): *syntactic markers* that apply to the term (e.g., "name"); *semantic markers* (e.g., "animal," "liquid," "metal," "time period"); a description of the *stereotype*; and a description of the *extension*.

By "stereotype" Putnam means a list of typical characteristics possessed to a large extent by each member of the term extension: for example, the stereotype associated with the term "tiger" will include characteristics such as having a striped coat, roaring, being fierce, not very tame, and so on, characteristics that tigers may have in whole or in part (nothing precludes a certain tiger being very tame and not roaring). This is that minimum amount of information that a speaker is obligatorily required to know in order to be part of the chain of communication related to a term, and also that little bit of the "psychological" that goes into the meaning of the term, given that that information can be considered as the concept related to the term, and with good approximation a concept can be considered something mental. But beware: the presence of such a mental element in Putnam's semantic conception does not make it a mentalistic conception. Indeed, the features in the list that constitute a stereotype do not amount to necessary and sufficient conditions for belonging to the extension of a word, and thus Putnam's fundamental point remains that what is in the head does not determine the extension and does not constitute the meaning of a word. Going with Kripke, Putnam argues that the description that can be given of the referents of a word is not "synonymous" with the word, and therefore does not exhaust the meaning of the word. If that were the case, every change in description (due, say, to an advance in the theory about those referents) would correspond to a change in the meaning of the term: if the description of the term "atom" changes over the centuries (as indeed it has), then "atom" changes its meaning from time to time, which seems counterintuitive. This is why Putnam has always reiterated, with a firmness that has never failed, that

> The meaning of "water" isn't fixed by a definition, either in terms of the observable properties of the substance or in terms of the properties mentioned in our latest scientific theory; it is fixed by the nature of our paradigms of water (Putnam 2016a: 208),

where the *paradigmatic* examples of water are all the water samples we can find around us. Therefore, the objects to which the term refers, the objects present in the natural environment play the main role in determining the meaning of a term of natural kind or physical magnitude. Here again is reaffirmed the place to which to trace the meanings of the terms studied by

Putnam: this place is *the world*, which in Putnam's science fiction example makes the word "water" as used on Earth and the word "water" as used on Twin Earth "homonyms, but not synonyms" (Putnam 2016a: 217).

EPISTEMOLOGICAL CONSEQUENCES

We have seen how and why Putnam arrives at the conclusion that for us water is H_2O in all possible worlds. It is the substance endowed with that particular chemical composition that we Earth speakers have always had to deal with, and it is this particular causal interaction that has reverberated on the meaning of the term "water": everything that counts as water for us is H_2O, and all that our term "water" can refer to are samples of H_2O. And this is true at all times, even in counterfactual situations in which we should find ourselves.

Kripke calls a linguistic expression a "*rigid designator* if in every possible world it designates the same object" (Kripke 1972: 48), granted of course that the object in question exists in the possible world under consideration. An expression such as "water" is thus a rigid designator. Drawing attention to the concept of rigid designator, and the consequences it has on a general philosophical level, constitutes the second debt incurred by Putnam to Kripke:

> What Kripke was the first to observe is that this theory of the meaning (or "use," or whatever) of the word "water" (and other natural-kind terms as well) has startling consequences for the theory of necessary truth (Putnam, 1975e: 232).

Before seeing these consequences it should be noted that Putnam has expressed the same thing that Kripke expressed with the concept of rigid designator by using the concept of *indexicality*: a term like "water" is an *indexical* term, since—just like indexical expressions such as "I," "this," "today," and the like—it changes denotation depending on the extralinguistic context considered ("I" changes denotation depending on who uses this term).

Let us now consider a statement in which a rigid designator appears: exploiting our usual term, let us take the statement "Water is H_2O." This is a statement that, if true,[19] is *necessary*—necessarily true, to be more precise—since there would be no possible situation in which it would be false: if true, it is so in all possible worlds, according to the usual definition of the metaphysical notion of necessity (Kripke points out that it is a "metaphysically necessary" statement). As is well known, this definition is coupled with the traditional definition of the epistemological notion of *a priori* (according to which a statement is a priori if, and only if, its truth value is known independently of experience), and with the traditional definition of the semantic notion of *analyticity* (according to which a statement is analytic if, and only if, its truth value depends solely on the meaning of the words that compose it,[20] and is therefore unrevisable unless the meanings of those words are revised: but then it would be another language). By virtue of such pairings all necessary statements are a priori; in addition, for some philosophers they are also analytic, while for still others, alongside the set of analytic necessary statements, there is a set of synthetic necessary statements. Very few in our philosophical tradition have therefore ever doubted that a necessary statement was a priori, even if it was synthetic a priori. Virtually no-one, that is, until Kripke and Putnam: to know the necessary truth expressed by the statement "Water is H_2O" requires a considerable amount of empirical research, and thus it is a genuinely *a posteriori* statement. Moreover, situations are easily imaginable in which we would deny that water is H_2O (while remaining true that it is): for example, we might mistakenly accept a chemical theory that leads to such a denial, or we might fail to notice some error in our laboratory analyses. We may therefore deny a necessary truth on entirely rational grounds.

FURTHER LINGUISTIC EXPRESSIONS

We have seen that Putnam's semantic conception does not apply to an entire natural-historical language, and that from the outset he himself delimits its scope to terms of natural kind and terms of physical magnitude. We have also noted, however, that there are many points of convergence between his and Kripke's analyses of proper names, which indicates how Putnam's conception can be extended to proper names as well. There is, however, the possibility of extending it in other directions too, a possibility suggested by Putnam himself. First, to the names of artifacts such as "chair," "bottle," and "pencil."[21]

Indeed, he points out that even these terms manifest a marked rigidity or indexicality, so much so that it would be wrong to assume that their meaning is entirely exhausted by a description of their characteristics: for example, of the characteristics encapsulated in the stereotype associated with them, which in the case of "pencil" might be something like "artifact mostly made of wood, enclosing a graphite lead, used for writing or drawing." If this were the entire meaning of the word "pencil," one could consider the listed characteristics (or possibly just a few) as necessary to be a pencil, the stereotype as a synonym for the word, and thus a statement like "A pencil is an artifact" as a priori and analytic. However—and here is another example of Putnam's science fiction imagination—it might turn out that pencils are not artifacts but living organisms: we might find that they deposit eggs, see these eggs hatch, and note the almost imperceptible growth of young pencils into adult pencils.

However strange and improbable, this is a perfectly conceivable situation that shows the mistake we would make if we considered the statement "A pencil is an artifact" analytic and a priori. The word "pencil," therefore, is not synonymous with any description.

A brief gloss on this example to avoid misunderstanding. If what is described in the example actually happened, should we conclude that pencils are organisms? Or should we say that there are no pencils? Similar to the case of "water," which of the two reactions is the correct one depends on what the world we live in is. The example just seen is equivalent to claim that in a possible world pencils are organisms; however, if this possible world is not the actual one, and if in the actual one pencils are artifacts, the statement "Pencils are artifacts" is for us Earthlings metaphysically necessary (in the sense of true in all possible worlds) and a posteriori (or epistemically contingent, as Kripke puts it). There is then no possible world in which pencils are not artifacts, and in the possible world described in the example pencils do not exist. Conversely, should that possible world coincide with the actual world, the statement "Pencils are organisms" is metaphysically necessary and epistemically contingent for us. Putnam's science fiction examples merely highlight how we behave and would behave verbally, what our referential intentions are and what our spontaneous verbal and non-verbal reactions are:

> when we use the word "pencil," we intend to refer to whatever has the same *nature* as the normal examples of the local pencils in the actual world. "Pencil" is just as *indexical* as "water" or "gold" (Putnam 1975e: 243).

Without developing it, Putnam finally gives a very sketchy indication about other expressions to which his semantic conception can be applied: verbs such as "grow" and adjectives such as "red," he argues, have indexical features. Presumably, what he means is that "grow" is to be interpreted as "developing in a certain natural environment," while "red" as "appearing in a certain way to beings endowed with a certain biological constitution." If, say, on TE the natural environment provides physicochemical conditions for a different development of living organisms and for a different refraction of light on the retina (which in turn might have developed under different conditions), then the meaning of "grow" and "red" would be different from the meaning these same terms have for us.[22]

FINAL GLOSS

We have seen that when Putnam came up with his own semantic account in the 1970s, he found congenial some of the results of the reflection on the reference of proper names that, independently, Kripke was advancing in those same years. And we have seen that among the consequences on a general philosophical level of their semantic reflection is the broadening of the concept of necessity to accommodate the notion of a posteriori necessity. It should be pointed out now that in the interpretation of the concept of necessity Putnam was following Kripke, and that this interpretation made this concept very strong. In line with it, in fact, a necessary statement is true (or false) in all possible worlds—in all of them, regardless of what happens in them. Thus interpreted, the concept was called "metaphysical necessity" by Kripke and "logical necessity" by Putnam, and—as we now know—identity statements such as "Water is H_2O" or "Temperature is mean molecular kinetic energy" fall under it. If these statements are true (and this can only be known a posteriori by virtue of the results of empirical research), then they are true in all possible worlds in which there is water or temperature, respectively. Kripke seasoned his support for the existence of metaphysically necessary statements by introducing the concept of *essence*—claiming that individuals referred to by names of natural kinds, physical magnitudes, or proper names possess a distinctive essence of their own. It is actually to this essence, he argued, that one refers when using these names: in all possible worlds in which a certain entity or individual exists, it is its essence that constitutes the denotation of the corresponding term, independently (to repeat) of anything that in a given possible world may happen differently from the actual world.

Some time later, however, since at least the 1980s, Putnam moved away as much from Kripke's essentialism as from the related notion of metaphysical necessity; with respect to the latter, indeed, he went so far as to say, "I now think that the question 'What is the necessary and sufficient condition for being water *in all possible worlds*?' makes no sense at all" (Putnam 1990b: 70), since it is not in this way that the identity of a certain substance can be ascertained. But why?

The answer is that the identity of a substance such as water depends on a physicochemical law: precisely the physicochemical law that applies in the present world—the world in which we happen to live. This is what makes us say that that liquid which on TE possesses all the phenomenological characteristics of water is actually not water but, at most, only plays the role of water. What Putnam realized a few years after the publication of "The Meaning of 'Meaning'" is that it is not true that "a criterion of substance-identity that handles Twin Earth cases will extend handily to 'possible worlds'" (Putnam 1990b: 69). One can, for example, easily imagine a possible world in which there is H_2O and in which, however, the chemical law governing the behavior of hydrogen and oxygen is such as to produce a different behavior of H_2O from what characterizes it on Earth: in such a situation we would not say at all that that liquid is water, even though it has the same physicochemical composition as our water. And yet the Kripkean criterion (which only looks at composition, i.e., the microstructure) leads us to conclude that that *is* water, thus revealing how the criterion has nonsensical consequences.

The fact is that within all possible worlds one can imagine anything happening, freely and without any restriction: precisely because one embraces the totality of the possible, anything and the opposite of anything can happen; and if a statement is true (or false) in this way—unconditionally—then one loses any basis for being able to say that a certain substance possesses a certain identity. That is why metaphysical necessity is a *strong* notion of necessity: it is a notion that characterizes a maximal degree of objectivity, or objective validity, a degree for which something can hold without any restrictions whatsoever, without any clause like "provided that . . ."—it just holds in every possible world. But such a maximal degree is completely unserviceable: on the basis of it, the notion of objectivity vanishes like smoke in the air. Therefore, the interpretation of necessity that for Putnam is suitable for dealing with the cases discussed here is the physical one: the statement "Water is H_2O" is *physically necessary*.

NOTES

1. A useful discussion of the tangle of elements involved in the externalism-internalism issue can be found in Chakraborty (2020), while an analysis of Putnam's causal theory with emphasis on context-sensitivity can be found in Chakraborty (2022).

2. Putnam takes particular care in delimiting his field of inquiry: first, he makes it clear that he will deal with the meaning of words, and not the meaning of sentences because, he argues, it is on the former that philosophical knowledge is most lacking; and second, that he will not consider *all* words but, as just mentioned, only those that refer to physical magnitudes (electricity, electron, temperature, etc.) and natural kinds (gold, tiger, rose, etc.).

3. Indeed, the extension of a term can be defined as the set of objects of which the term is *true*.

4. Later Putnam explained the point as follows: "For example, according to Plato, ideas are extra-mental entities, but knowledge of them is supposed to be innate in the mind, and capable of being brought to consciousness by an act of 'recollection.' For Aristotle, they are both mental and extra—mental. The very same 'idea' or 'form' is supposed to be capable of existing in things, but also, minus its matter, in the mind. In the middle ages, Conceptualist and Nominalist views were added to the Platonic and Aristotelian alternatives. For the Conceptualists ideas are explicitly mental entities; for the Nominalists they are, of course, 'names,' but the understanding of 'names' is supposed to lie in each individual mind. In the modern period, for an empiricist like Hume, ideas are hardly distinguished from mental images, and are certainly in the mind. Thus, in spite of the variety of metaphysical theories about the nature of

concepts, this much was not doubted: concepts were uniformly thought of as capable of being completely contained in or recollected by 'the mind' (which was itself conceived of as a private theater, isolated from other individuals and from the 'external world')" (Putnam 1996: xv).

5. Cf.: "In *De interpretatione* [Aristotle] laid out a scheme which has proved remarkably robust. According to this scheme, when we understand a word or any other 'sign,' we associate that word with a 'concept.' This concept determines what the word refers to" (Putnam 1988: 19).

6. "Historically speaking, (I) represents the temptations of empiricism and psychologism" (Floyd 2005: 21).

7. The opposite, on the other hand, is well established: two terms can have the same extension and different intension, as the case of "creature with a heart" and "creature with a kidney" would stand to prove.

8. "Historically speaking, [. . .] (II) represents the temptations of rationalism and logicism" (Floyd 2005: 21).

9. The psychological state referred to in principle (I) could be described as 'knowing that intension *I* is the meaning of term *t*'; this state thus determines the intension of *t* and, by principle (II), its extension.

10. "That psychological state does not determine extension will now be shown with the aid of a little science-fiction" (Putnam 1975e: 223). "With that humble sentence in 1975 Hilary Putnam changed the face of philosophy forever. Twin Earth burst on the scene like the legendary meteor that did in the dinosaurs; it has been reverberating through philosophy ever since. With implications stretching far beyond its original domain in the philosophy of language and philosophy of psychology, it has left almost no area of contemporary analytic philosophy untouched" (Pessin & Goldberg 1996: xi).

11. In responding to the objection that, this being the case, we did not know the meaning of the word "water" until we developed modern chemistry, Putnam notes that "knowing the meaning of a word" can be understood in at least three ways: (*a*) knowing how to translate it, (*b*) knowing what it refers to, in the sense of possessing the ability to establish its denotation independently of the mere use of the word, and (*c*) possessing tacit knowledge of its meaning, in the sense of being able to use the word in the various contexts in which it may be used. "The only sense in which the average speaker of the language 'knows the meaning' of most of his words is (*c*). In that sense, it was true in 1750 that Earth English speakers knew the meaning of the word 'water' and it was true in 1750 that Twin Earth English speakers knew the meaning of their word 'water.' 'Knowing the meaning' in this sense isn't literally knowing some *fact*" (Putnam 1988: 32).

12. A usual objection raised against this example is that even in the case of average speakers—those who cannot tell an elm from a beech tree—their mental representations of an elm and a beech must be different since they *know* that elms and beeches are different species. However, the objection is fallacious. It is certainly true that I know that they are two different species and that, therefore, my mental representation of an elm includes that it is not a beech, and vice versa for my mental representation of a beech. But this means nothing more than "my mental representation of an elm

includes the fact that there *are* characteristics which distinguish it from a beech, and
my mental representation of a beech includes the fact that there *are* characteristics
which distinguish it from an elm. The situation is totally symmetrical [. . .] Apart from
the differences in the phonetic shapes of the names (which, as we have seen, cannot
be a part of the meaning of the names), there is no difference between my 'mental representation' of a beech and my 'mental representation' of an elm" (Putnam 1988: 29).

13. Juliet Floyd correctly points out that this "unforgettable externalist slogan [. . .]
can mislead, however much truth there is in it. For the notions of something's being
'in the *head*' and 'in the mind' are considerably more problematic than the slogan
suggests. Semantic externalism, as Putnam now presents it [. . .] entails that word-
(and sentence-) meanings are best not conceived of as entities of which we could
sensibly ask, 'Where then *are* they (if not in the *head*)?' The point, then, is that we
should stop trying to conceive of them as objects that either do or do not measure up
to 'truly scientific' scrutiny" (Floyd 2005: 25). As we will see in the following chapters, the ceaseless rethinking to which Putnam subjected his own externalism was to
lead him to a better conception of what the human mind is. Cf.: "Of course, denying
that meanings are in the head must have consequences for the philosophy of mind,
but at the time I wrote those words I was unsure as to just what those consequences
were" (Putnam 1996: xvii).

14. An interesting analysis of Putnam's notion of the division of linguistic labor
can be found in Wagner (2020).

15. Putnam clarified that Kripke's account "was developed independently of my
own. My own was first presented in lectures at Harvard in 1967–1968, and at lectures
at Seattle and at the University of Minnesota the following summer; neither Kripke
nor I published our accounts until a few years later" (Putnam 1988: 130); in fact, "my
first publication with an externalist and anti-individualist account of reference was
'Is Semantics Possible?' [Putnam 1970], published the year Kripke gave the lectures
collected in his *Naming and Necessity*" (Putnam 2016c: 130). Kripke, for his part,
stated that "the ideas in *Naming and Necessity* evolved in the early Sixties—most of
the views were formulated in about 1963–64" (Kripke 1972: 3).

16. "Kripke's work has come to me second hand; even so, I owe him a large debt
for suggesting the idea of causal chains as the mechanism of reference" (Putnam
1973: 198).

17. Putnam has repeatedly pointed out that this does not amount to saying that the
reference of a natural kind term such as "water" is fixed exclusively by a scientific
theory: what is established in ordinary contexts by ordinary speakers can also be
water, where central is the aforementioned empirical assumption with all the interest-
relativity and context-sensitivity that it carries with it (cf. Putnam 2015b: 80). The
"environment" in question here is therefore not only that which can be described in
physical-natural terms. In other words, the obvious difference between ordinary usage
and scientific usage should not lead to the conclusion that the latter is privileged. Far
from any scientistic reductionism, the fact is that "ordinary language and scientific
language are different but *interdependent*" (Putnam 1986: 282). Quite simply, what
happens is "that there are causal constraints on reference, not that the referring *is* the
causal connection" (Putnam 1992b: 165).

18. Having stressed "the contextual and contingently situated complexity involved in our human ways of [. . .] weighting components of the meaning-vector," Juliet Floyd maintains that this contextual complexity "was what allowed Putnam to soften the notion of *determination* at work in principle (II) to an acceptable one" (Floyd 2005: 31). There can be no *systematic* explanation of such a "determination."

19. The specification "if true" is very important here: it implicitly emphasizes that it is scientific research that must determine the truth or falsity of such a statement, and that therefore this truth/falsity depends on the progress of scientific research—which is virtually endless. Even from the perspective of the current state of knowledge, the statement in question is not precisely true: "Normal water is actually a quantum mechanical superposition of H2O, H4O2, H6O3 . . . Very little (if any) water is simply H2O" (Putnam 2016a: 207).

20. This, as we saw in Chapter 1, is the definition of analyticity offered by Quine, which is more general than the Kantian definition.

21. For a criticism to the extension of Putnam's analysis to artifact terms, cf. Schwartz (1978).

22. In Burge (1979) Tyler Burge discusses favorably the possibility of such extensions of Putnam's analysis.

Chapter Five

The Philosophy of Science

We are now in a position to appreciate the implications of Putnam's semantic externalism for the philosophy of science, particularly the way its two central features, the social and the ontological, are intertwined—and much more than "The Meaning of 'Meaning'" implied. As Gary Ebbs has pointed out, "the interdependence of our ontological notions and our substantive beliefs is implicit in Putnam's early arguments against logical positivism" (Ebbs 1992: 2), where these beliefs are guided by community-shared norms that, in turn, are inherently open-ended and not given once and for all. A proper understanding of the norms underlying our rational inquiries is indeed the task Putnam never ceased to set himself.

AGAINST LOGICAL POSITIVISM

As we have already had occasion to see, Putnam matured his philosophical thought in opposition to logical positivism, a movement of thought that, in his formative years, constituted much of the general cultural background.[1] Admittedly, logical positivism was a composite and non-uniform movement, enlivened by the contributions of eminent thinkers who were not always in agreement with each other and characterized by an incessant and uncommon internal dialectic that came to invest its own theoretical cornerstones—cornerstones that ended up being abandoned because of honest self-criticism and not only because of criticism from outside. Decisive among the latter were those of authors such as Willard Van Orman Quine, Karl Popper, Thomas Kuhn, Paul Feyerabend, and Putnam himself.

Now, while keeping in mind the diversity among the positions of logical positivists, it is possible to say that their empiricist approach leads, from a metaphysical point of view, to idealistic outcomes. Putnam identifies the reason for this: "the idealistic element in contemporary positivism enters precisely through the theory of meaning" (Putnam 1973: 207), that is, through the well-known *principle of verification* according to which the meaning of a statement is the method of its empirical verification. Although gradually weakened because of the difficulties encountered in its application, this principle establishes that only meaningful propositions, that is, those that can be empirically verified (at least in principle), have cognitive value: these are the propositions of which it can be said that they describe reality, in the sense that the objects and situations to which they refer are to be considered as actually existing. It follows then that only that of which one can have actual or potential experience counts as *real*. Therefore, if it is solely by means of experience that we are given the world, and if this is done by virtue of the acquisition of a set of sense data by the human perceptual apparatus, it is reasonable to assume that the world depends on this apparatus and the ways in which it encodes sense data, unless we have a sound argument intended to show that beyond the sense data there is a world external to the perceptual apparatus and to some extent independent of it; but an argument of this kind that was *not* itself linked to the plane of experience would have the characteristics of the type of metaphysics that logical positivists branded—especially in their early phase—as bad philosophy devoid of cognitive content. Therefore, one cannot but conclude that from a (logical) empiricist perspective, the world turns out to depend on us, that is, on the ideas, beliefs, and knowledge that inform the process of encoding sense data. Empiricism in general, and its twentieth-century version—logical positivism—in particular, thus show that they are marked by a distinctly idealistic metaphysical bent.

Of course, deciding in the abstract what is the most correct metaphysical attitude has always been a very thorny problem—one of the thorniest problems in philosophy. However, the philosopher of science has an important opportunity: she must show that whatever metaphysical framework she places her reasoning within, that framework must be compatible with *the actual practice of scientists*, with what they do when they try to uncover the secrets of nature. This is what Putnam calls the *adequacy claim*, that is, the claim that one "is entitled to accept standard scientific theory and practice" (Putnam 1973: 207). From this point of view, (logical) empiricists therefore have the task of showing that their general metaphysical and epistemological perspective, informed by the principle of verification, faithfully portrays the way scientists intend, albeit sometimes implicitly, their cognitive enterprise—an enterprise that can be summed up in the effort to develop theories and apply them to reality. To begin with, a belief that would seem to underlie the scientists' work—a commonsense belief, therefore prior to any mature philosophical position—is that nature offers them a "material" reality. Are logical empiricists, then, able to show that the material reality with which scientists deal can be treated from an empirical point of view?

OBSERVATION LANGUAGE AND THEORETICAL LANGUAGE

One way to show this—a way, ultimately, to point out that there is compatibility between the materialist perspective and their tendentially idealist perspective—is to show that both scientific language and commonsense language (languages called *thing-languages* because they are both centered on objects and their actual existence) are translatable into *phenomenalistic language*—centered on sense data. However, no (logical) empiricist has ever provided such a proof: not Berkeley, not Mach, and not even Carnap—the only one to set the goal of translating thing-language into phenomenalistic language on a precise philosophical basis—who only succeeded in providing the preliminaries of such a translation (cf. Carnap 1928): "In short, no demonstration is given at all that the positivist is entitled to quantify over (or refer to) material things" (Putnam 1973: 208), and, after all, the logical positivists themselves ended up realizing that the translation they were seeking was unrealizable.

Carnap himself inaugurated a new phase, characterized by the substitution of phenomenalistic language with "observation" language as the epistemologically and semantically legitimate basis to which to reduce the *unobservable statements* of the language of scientists—those in which "theoretical terms" appear, terms that do not describe observable entities but rather entities postulated for the explanatory and predictive needs of the theory in which these terms occur (cf. Carnap 1936–37, 1956). However, even the phase inaugurated by Carnap failed to dispel the suspicion that the logical positivist philosophy of science is incapable of providing a plausible account of scientists' practice.

As a result of the distinction between observation and theoretical language, logical positivists regarded scientific theories as "partially interpreted calculi," that is, axiomatic systems that possess an interpretation only relative to their observation terms; the other part of the vocabulary of theories, that consisting of theoretical terms, somehow deduces its meaning from the observation terms themselves. Thus, there is a double distinction in the vocabulary of scientific theories between, on the one hand, observation terms (referring to publicly observable objects and their publicly observable properties) and observation statements (containing only observation terms and logical vocabulary), and, on the other hand, theoretical terms (referring to unobservable objects and properties such as electron, gene, Super-Ego, conductivity) and theoretical statements (containing theoretical terms). It should be borne in mind that only the first horn of both distinctions possesses empirical meaning for logical positivists—the only meaning that according to them is worthy of the name, the one endowed with cognitive value—and is therefore directly interpreted.

On this double distinction and on the notion of "partial interpretation," Putnam's critique is firm and decisive: "if an 'observation term' is one that cannot apply to an unobservable, then there are no observation terms" (Putnam 1962c: 217). On closer inspection, in fact, all terms can be applied to unobservable objects: Newton himself, for example, used the predicate "red" in this way when he hypothesized that red light consisted of red corpuscles; and, moreover, it is well known that even a three-year-old child can understand stories about "people too little to be seen"—a phrase, the latter, in which only observation terms appear. As for theoretical terms, it is their very definition as designating something unobservable that is for Putnam misleading: it would follow that terms such as "angry," "loves," and so on would be theoretical simply because they do not designate something publicly observable. By contrast, if we adopt an obvious and innocuous definition of theoretical term, such as "term derived from a scientific theory," it will turn out quite naturally that "satellite" is a theoretical term even if it refers to observable objects.

Similarly for the other distinction, that between observation and theoretical statements. That the former can contain theoretical terms can be easily ascertained by noting how widespread are such statements as "We also *observed* the creation of two electron-positron pairs" (Putnam 1962c: 219); conversely, it can be said that even the main case of a theoretical statement—a scientific theory—can refer only to observables: an example of this is "Darwin's theory of evolution, as originally put forward" (Putnam 1962c: 217).

Both the distinction between observation and theoretical terms and the distinction between observation and theoretical statements are therefore very blurred and unclear. But then one can hardly support a conception of scientific theories according to which "theoretical terms are only partially interpreted, whereas observation terms are completely interpreted, if no sharp line exists between the two classes" (Putnam 1962c: 220). And not only that: the very idea of "partial interpretation of theoretical terms," on close examination, seems completely unjustified. Why should theoretical terms be understood only partially, as if they had an *imprecise* meaning? And why, assuming they have it, can they be understood only through observation terms? That it can only be the latter that give meaning, albeit partially, to theoretical terms is, in Putnam's view, entirely misleading: for it is not in this way that theoretical terms acquire and retain the meaning they have, for the reasons we saw in Chapter 4. Indeed, the causal semantic conception developed by Putnam highlights how theoretical terms have a status of their own that is not reducible to that of observation terms. To disavow this is to debase not only the role of theoretical terms but also that of the theories containing them. A brief explanation of this last point will bring out with conclusive clarity how the idealistic background of logical positivism prevents a faithful description of the practice of scientists.

THE LOGICAL POSITIVIST DEBASEMENT OF SCIENTIFIC THEORIES

We have said that in the wake of Carnap the logical positivists regarded theories as partially interpreted calculi, and we have seen what the "partiality" of their interpretation consisted of. But, one might ask, why "calculi"? The use of such an appellation derives from the belief that a theory is nothing more than a *tool* for obtaining explanations of phenomena that have happened and predictions of future phenomena, and this belief in turn reveals the deep distrust that the logical positivists had of the concept of truth, which they judged to be hopelessly "metaphysical" (in their typical derogatory meaning of this adjective: something lacking cognitive significance). Indeed, in itself the concept of truth has nothing that can be traced back to experience, and therefore, as it is, it cannot be accepted by a logical positivist. In particular, it cannot be used as a criterion for accepting scientific theories; if, however, theories are viewed as tools, then at most it can be said of them that they are either useful or useless, not that they are (approximately) true or false. But this is enough to rule out a logical positivist's offering an adequate account of scientific practice:

> When a realistically minded scientist—that is to say, a scientist *whose practice* is realistic, not one whose official "philosophy of science" is realistic—accepts a theory, he accepts it as true (or probably true, or approximately-true, or probably approximately-true). Since he also accepts *logic* he knows that certain moves *preserve truth* (Putnam 1973: 210).

In case he accepts a theory T_1 as true, and later a theory T_2 as true, then he knows by purely logical means that the theory that results from the conjunction T_1 & T_2, let us call it T_3, is also true, and will therefore accept T_3. Such behavior, despite being one

> of the simplest moves that scientists daily make, a move they make as a matter of propositional logic, a move which is central if scientific research is to have

any *cumulative* character at all, is totally *arbitrary* if positivist philosophy of science is right (Putnam 1973: 210).

The reason for such arbitrariness is quickly stated: for the positivist, to accept T_1 as true is to accept a useful calculus insofar as it allows for successful predictions and, likewise, to accept T_2 as true is to accept a useful calculus insofar as it allows for successful predictions; but it does not follow from this at all that the conjunction of the two theories leads to successful predictions and can therefore prove useful.

In fact, a "successful prediction" is nothing more than a statement that establishes the occurrence of a certain phenomenon under certain circumstances and that is verified empirically (a sign that the prediction was correct). Now, both predictions and verification procedures depend on the conceptual apparatus of a theory, and can therefore vary from theory to theory. They are *intratheoretical*, and equally intratheoretical is the predicate "leads to successful predictions" which, for the positivist, takes the place of the predicate "is true." But, as we have seen, the combination of two theories forms a new and different theory, T_3, and this one will establish its own use of the predicate "leads to successful predictions," and nothing guarantees that, necessarily, the results of such use will conform to those obtained separately from T_1 and T_2, and thus that the combination of the conceptual apparatuses of T_1 and T_2, and their related verification procedures, will retain their predictive power. And the reason according to Putnam is simple: the predicate that for the logical positivists plays the role of truth—the predicate "leads to successful predictions" or, equivalently, "empirically verifiable," or "assertable because of the warranty offered by the best current theory"—"does not have the *properties* of truth" (Putnam 1973: 211), for example, logical properties such as that of the conjunction exemplified above.

THE IDEALISTIC FALLACY

This is an important statement. It rests on an assessment of the *nature* of truth from which its special status stands out—a status that clearly distinguishes the concept of truth from any other concept. It is by virtue of this nature that to eliminate truth in favor of some other concept would for Putnam be tantamount to committing a fallacy: the *idealistic fallacy*. Indeed, since

> for any predicate P the idealist may want to substitute for "true" one can find a statement S such that [the conclusion] "S might have property P and still not be *true*" follows from our causal theory of learning (Putnam 1978a: 108–109),

truth turns out to be irreducible to other concepts, so that the logical positivist attempt to replace it with a concept that is more acceptable from an empiricist point of view inevitably turns out to be fallacious: any such substitution will never entirely capture the content of truth.

This is why from the logical positivist perspective there follows a debasement of theoretical terms and theories, a debasement we now understand to depend on the refusal to recognize truth as having full and autonomous consistency. For the logical positivist, empirically verified statements are true; but this is a narrow use of the word "true," reflecting an equally narrow conception of truth. This use of "true" as a synonym for "verified" or "verifiable" or "assertible" allows at most to say that all *observation* statements that are theorems of T_1 are true, not that *all* statements that are theorems of T_1 are true, because statements derivable from a theory that are unverified or unverifiable possess no cognitive value for the logical positivist. It follows that a theory as a whole can never be said to be true (in the logical positivist sense), because not all of its theorems are verifiable. But this seems too forced and counterintuitive a conclusion to be acceptable: on the one hand, it makes a scientific theory an unjustifiably one-sided system, "unbalanced" toward its observation components, whereas instead observation and theoretical components can only complement each other; on the other hand, the logical positivist interpretation of truth is strictly intratheoretical, just like the verification procedures on which it is made to depend, whereas truth can only be untied from any theory if it is to form a basis on which to evaluate the theories themselves (particularly those competing with each other).

In conclusion,

> the positivist may teach in his philosophy seminar that acceptance of a scientific theory is acceptance of it as "simple and leading to true predictions," and then go out and do science (or his students may go out and do science) by verifying theories T_1 and T_2, conjoining theories which have been previously verified, etc.—but then there is just as great a discrepancy between what he teaches in his philosophy seminar and his *practice* as there was between Berkeley's teaching that the world consisted of spirits and their ideas and continuing in practice to daily rely on the material object conceptual system (Putnam 1973: 211).

If, then, what scientists concretely do (the way, often implicit, they understand their theories, the relation of these theories to reality, and the role reality plays in legitimizing some theories and delegitimizing others) constitutes a kind of test case for the plausibility of philosophical conceptions, then the inability of positivists to faithfully account for the scientists' practice condemns *ipso facto* their philosophical approach, and in particular their distinctive conception of what a theory is. Indeed, from Putnam's critique it emerges that a scientific theory is a predictive and explanatory endeavor in which all its components are concerned: not only the observational but also the theoretical. It emerges that a theory is not a mere predictive and explanatory tool of which, at most, one can say that it is "useful," "leads to true predictions," "is assertable with guarantee" (all expressions that the positivist would substitute for "true"); in other words, it is not a mere linguistic tool whose relevant parts are, from a semantic and epistemological point of view, only the observational ones—the theoretical ones representing a philosophical dead weight to be hopefully reduced to the former. On the contrary, it is as an integrated set of terms and statements that a scientific theory is *accountable* to reality—an autonomous and independent reality that determines whether the theory is true or false.

Note how Putnam closely links the failure of positivism to account for what scientists do with the positivist failure to interpret the notion of truth, as well as the closely related notion of reference. And since both notions merely point to the relationship between language and reality, it follows that part of the failure of the (logical) positivist philosophical perspective is due to the general metaphysical background which, as we have seen, displays a distinctly idealistic bent. It is the critique of idealism in whatever form it may present itself that constitutes one of the cornerstones of Putnam's thought, a point never abandoned even through the changes to which—as we shall see in Chapters 6 and 7—Putnam subjected his own metaphysical position. The point is already very clear in the early years of his reflection. That scientists hold an implicit realist view is for Putnam a fact, something that makes *realism science's philosophy of science* (as during a discussion a colleague and friend of his, Rogers Albritton, had to state):

> The realist, in effect, argues that science, should be taken at "face value"— without philosophical reinterpretation—in the light of the failure of all serious programs of philosophical reinterpretation of science, and that science taken at "face value," *implies* realism (Putnam 1978a: 37).

To think otherwise would make the success of science a *miracle*. But why?

SCIENTIFIC REALISM

The reason is implicit in the notion of a theoretical entity, which, as we saw above, is a cornerstone of the logical positivist interpretation of science. For if the unobservable entities to which scientists refer are regarded only as "theoretical conveniences," that is, entities postulated solely for the purpose of not overcomplicating a theory, and thus ultimately as non-existent,

> then it is a *miracle* that a theory which speaks of gravitational action at a distance successfully predicts phenomena; it is a *miracle* that a theory which speaks of curved space-time successfully predicts phenomena (Putnam 1978a: 19; see also Putnam 1975e: 237; 1975f: 73; 2012c: 55; 2012f: 91ff).[2]

From this it follows that the option that best satisfies the *adequacy claim* cited at the beginning of the chapter is to regard the theoretical entities that scientists talk about as actually existing and belonging to a world independent of human cognitive efforts. This is the position usually called *scientific realism*, a position Putnam subscribed to from the very beginning of his career, provided it is not regarded as incorporating the thesis that "*all* knowledge worthy of the name is part of 'science'" (Putnam 1978a: 20)—an anti-reductionist position on which he would never cease to insist.[3] And throughout his career he rested his interpretation of scientific realism on two principles formulated in the 1970s by Richard Boyd and summarized as follows:

(1) Terms in a mature science typically *refer*.
(2) The laws of a theory belonging to a mature science are typically approximately *true* (Putnam 1978a: 20; see also Putnam 1975c: 290; 1975f: 73).

These two principles have a realist character because they regard terms and laws as representing things and magnitudes that are real and independent of the theories within which those terms and laws are formulated, making semantic notions of reference and truth equally independent. By virtue of this independence, truth and reference take on a trans-theoretical significance that proves useful when one has to explain the success of science.

Indeed, according to Putnam, it is only through these principles that a good explanation of the *progress* of scientific knowledge can be obtained—that is, of the way scientific theories succeed one another according to an order in which the one that comes later is "better" than the previous one. For suppose that a scientist believes in the two Boydian principles and is looking for a theory T_2 that is better than a long-accepted theory T_1. According to the second principle, T_1's laws are approximately true: from this it follows that T_2 will have to take this fact into account, will have to have the property that T_1's laws are approximately true from T_2's point of view, and thus contain it as a limiting case. The result is that the range of possible candidates to be T_2

narrows considerably, thus increasing the possibility of success. According to the first Boydian principle, then, the terms of T_1 have a reference: T_2 will then have to have the property that from its point of view referents can be assigned to the terms of T_1, and this again will reduce the number of possible theories with which to replace T_1. Since the notions of reference and truth that the realist postulates are trans-theoretical, they can help build a bridge from one theory to another by allowing theories to gradually *converge* on their cognitive target.[4]

Note how all this tallies with the causal conception of meaning examined in Chapter 4—the semantic externalism according to which the meaning of many of our words, including the words for physical magnitude used by scientists, is primarily given by entities existing in the world, and therefore "ain't in our heads." Indeed, Putnam's causal conception offers a simple and elegant way of explaining how theoretical terms keep their references constant through the development of theories over time, thus ensuring commensurability between the uses of the same term made by scientists working within different conceptual frameworks—or between the uses that the same scientist may make at different stages of his or her research: for example, Niels Bohr's use of the term "electron" around 1900 and its use when, a few decades later, he participated in the birth of quantum physics (cf. Putnam 1973: 197; 1975c: 275; 1978a: 24). An undoubted advantage of Putnam's semantic conception is thus that it harmoniously integrates the account of the progress of science just seen: scientists living in different cultural epochs (or a single scientist considered at two different stages of his or her activity) refer to the same thing only if they are causally related to the same thing, and science progresses as the theories they develop succeed each other as better descriptions of the same entities.

Conversely, it is precisely an incommensurability between different uses of the same terms that stems from the logical positivist idea that "terms in scientific theories [. . .] have their meanings fixed by 'definitions'": indeed, it follows from this idea that "whenever a scientific revolution forced us to revise those 'definitions'—something [logical positivists] recognized had already happened more than once—it followed from their account that the *reference* of those terms changed" (Putnam 2016a: 201). And this is because, as the theories within which definitions are formulated change, so do the definitions themselves and thus the meanings of all terms.

It is worth mentioning here a methodological principle involved by Putnam in his conception of meaning, because its observance helps to establish the correct reference of proper names, physical magnitude terms, and natural kinds terms: the *principle of the benefit of the doubt*. We have seen in Chapter 4 that to ensure a link between the uses of a term and the entity to which it refers there is a causal chain of transmission whose links represent the uses that different speakers make of the term, where it is of the utmost importance that each speaker's use is governed by the referential intention to refer to the entity to which refers the use of the speaker from whom the use was causally learned. Precisely the indispensable presence of referential intentions makes it clear that once a referential chain has been specified, one has done only, so to speak, half the work: in addition to that one must be able to determine the concrete meanings of the words used by a linguistic community. Whether that community is the one to which we ourselves belong, or whether it is different from our own, we are left with the task of interpreting the verbal expressions of other speakers by attributing meaning to the parts that make up those expressions. The principle of the benefit of the doubt assumes central importance in this very interpretive task. It validates our assuming that the interpreted speaker would accept a reasonable reformulation of his or her own statements in the event that he or she were made aware of the acquisition of new knowledge about the object of those statements—exactly as Bohr did in the above example. In short, when we give the benefit of the doubt to a speaker who is using a certain term—a speaker who, implicitly or explicitly, shows that he or she holds certain beliefs about the referent of the term—we take the liberty of doubting those beliefs, given the likelihood of having to replace them in whole or in part with better beliefs about the same object. Putnam argues that, "like all methodological principles it is partly a descriptive principle" (Putnam 1975c: 275), since

> it is a fact that we can assign a referent to "gravitational field" in Newtonian theory *from the standpoint of* relativity theory (though not to "ether" or "phlogiston"); a referent to Mendel's "gene" from the standpoint of present-day molecular biology; and a referent to Dalton's "atom" from the standpoint of quantum mechanics (Putnam 1978a: 22).

And yet, he continues, it is also "a *normative* principle: we *should* honor it, for otherwise stable reference to theoretical entities would almost surely be impossible" (Putnam 1975c: 275).

THE MANY FACES OF REALISM

The position that best satisfies the *adequacy claim* is therefore the realist one which, in the context addressed in this chapter, takes on the guise of scientific realism. This kind of realism, Putnam argues, is a kind of "empirical" theory: first, it is embedded in our system of scientific knowledge, a thesis expressed by the claim that it is science's philosophy of science; second, it has many of the characteristics of an empirical theory—in particular, it explains *facts*.

One of the facts that this theory explains is the fact that scientific theories tend to "converge" in the sense that earlier theories are, very often, limiting cases of later theories (which is why it is possible to regard theoretical terms as preserving their reference across most changes of theory). Another of the facts it explains is the more mundane fact that language-using contributes to getting our goals, achieving satisfaction, or what have you (Putnam 1978a: 123).

Therefore, to the extent that it is an empirical hypothesis (or, rather, is "analogous" to an empirical hypothesis) realism stands or falls depending on what the facts say—in other words, depending on whether or not it succeeds in having an effective explanatory and predictive charge concerning the facts it is supposed to explain and predict. And Putnam never wavered in his full faith in this charge.

But scientific realism is but one of the faces that realism can take, as Putnam has masterfully shown and as we shall see in the next two chapters.

NOTES

1. For example, the exchange Putnam had with Hans Reichenbach and Rudolf Carnap in the early 1950s was very fruitful.

2. For some references on the enormous discussion triggered by Putnam's thesis, cf. Boyd (1991); Alai (2012), (2014), (2017), (2020); Psillos (1999), (2001); Lipton (2001); Hoyningen-Huene (2011); Lee (2014); Menke (2014).

3. Cf.: "there is a tradition with which I would not wish to be identified, which would say that scientific knowledge is all of man's knowledge. I do not believe that ethical statements are expressions of scientific knowledge; but neither do I agree they are not knowledge at all. The idea that the concepts of truth, falsity, explanation, and even understanding are all concepts which belong exclusively to science seems to me to be a perversion" (Putnam 1975a: xiii).

4. It is actually using the concept of "convergence" that scientific realists usually explain the progress of scientific knowledge.

Chapter Six

The Faces of Realism and Truth

The metaphysical backdrop against which Putnam's philosophical reflection stands out is avowedly realist, and this is most evident with regard to the causal conception of meaning. The core of the meaning of a term such as "water" consists of the chemical compound that actually exists in nature, the one to which all uses of the term are causally related, that is causally responsible for certain effects, and that exists quite independently of the mind and language of human beings. The general implication of semantic externalism for metaphysics is, thus, realism. But there is more than one way to be a realist and, indeed, the quest for the proper way to be a realist assumed central prominence at a certain point in Putnam's career (roughly, beginning in the second half of the 1970s) and did not leave him thereafter. Several stages he went through in what he himself later had to call "a long journey" (Putnam 1999: 49; cf. also Baghramian 2008a), and it can be said that each stage represents an attempt to model a convincing naturalist perspective. This chapter and the following one add to the mosaic we have outlined so far the pieces that characterize the particular variety of naturalism offered by Putnam. Let us therefore follow the steps of that long journey.

METAPHYSICAL REALISM

The aforementioned feature of independence helps to understand what it means to be a realist: it means holding that the reality with which we deal daily, and on many levels, is not a product of the human mind or a construction of our language. On the contrary, aside from the instances of active human intervention in reality, the latter is in its broad outlines something autonomous and disengaged from our intellectual and practical activity. It follows that the innumerable cognitive attempts in which such activity takes place cannot have any epistemic guarantee capable of ensuring its success. Such a guarantee could arise only from the existence of a pre-established link between human cognitive faculties and the world, a link that a realist position tends to deny. We can therefore say that realism is generally an expression of a *non-epistemic* conception of reality, that is, a conception in which human knowledge plays a limited role.

The same can be said of truth, and it is no accident that there is a kind of parallelism between the concept of truth and that of reality. Whichever interpretation of the concept of truth we find most pertinent, it is from the statements that are true that an image of reality arises. A realist's conception of truth is therefore non-epistemic too—the truth of our statements, just like reality, has no guaranteed connection with human knowledge. It is independent of it. All of this seems to act in the background of Putnam's reflection on language, science, mind, and mathematics that we have seen in the previous chapters.

Four Assumptions

The option in favor of realism that Putnam had left in the background of his early philosophical reflection begins to gain prominence as he became aware of the fact that "the major problem of philosophy [is] the problem of the way language and thought 'hook on' to the world" (Putnam 1983b: 315; see also 1985: 43, 1989c: 105, and 2012c: 58), i.e. the problem of *intentionality*. For reasons due to the influence he received from both Michael Dummett and the classical pragmatists, particularly William James and John Dewey, at the turn of the early 1980s Putnam developed a close critique of his own early realist position—the one that had been judged to need no special clarification and had thus been left in the background. In abandoning this position he seeks to identify it precisely: he gives it the name "metaphysical realism," judges it to be almost identical to the transcendental realism of which Immanuel Kant spoke, and considers it to be characterized by four assumptions, namely the existence of:

(1) a fixed totality of all objects;
(2) a fixed totality of all properties;
(3) a sharp line between properties we "discover" in the world and properties we "project" onto the world;
(4) a fixed relation of "correspondence" in terms of which truth is supposed to be defined (Putnam 1994e: 183).

As can be seen, from these assumptions we derive the image of a *static* reality, endowed with a rigid, independent structure, *ready-made*[1] and waiting to be probed by the cognitive activity of the human mind. A reality that in Kantian terms is made up of "things-in-themselves," and that can easily (though not necessarily) be conceived of as woven out of *essences*, that is, objects endowed with an intrinsic nature that possess equally intrinsic properties: those which, as the third assumption dictates, it is possible to "discover." Hence the dichotomy between these properties and the properties that we instead "project" onto objects—properties such as colors, smells, etc.

One derives, in other words, a picture of the world and the human mind as two clearly separate and distinct entities that are brought into connection when the cognitive efforts of the human mind are successful and manage to formulate true statements, those statements linked by a correspondence relationship to the portion of the world on which they focus.

Note that this relationship is conceived as *unique*, in the sense that it would always be the same kind of relationship regardless of the type of statement considered: it follows that, although at an intuitive level one can understand in what sense a statement about physical reality is true, considerable difficulties arise when it comes to understanding how, on this basis, a mathematical statement or a moral one can be true.

And note also that the above image can only be effectively "surveyed" from a non-human perspective: from a point of view that is external to both the world and the mind (and the language it employs) and that can embrace the whole scene at once. It is an image that presupposes the so-called "God's Eye point of view," or a "cosmic exile" according to Quine's expression, or a "view from nowhere" as Thomas Nagel put it, or the "spectator's theory of knowledge" as Dewey critically called it, or "an Archimedean point of view" according to a traditional image also used by Descartes and Hegel.[2] It is, to finish with metaphors, an image that for Theodor Adorno requires "a metaphysics from the peephole."

Brains in a Vat

The non-epistemic conception of reality and truth represented by metaphysical realism may run into a well-known, and seemingly unsolvable, epistemological problem: the case of the so-called *brains in a vat* (BIV). It is nothing more than a contemporary version of the skeptical hypothesis of the evil genius presented by Descartes in the *Méditations philosophiques*, a version that is due to Putnam himself (1981: ch. 1) and has been discussed by various authors since the 1980s.[3] Suppose that unbeknownst to us, a very cutting-edge but also very insane neuroscientist manages to anesthetize us, extract our brains from our skulls, and keep them alive by placing them in a vat full of nutrients. Let us now imagine that, by means of a dense network of wires, he manages to connect the myriad synapses of our brains to an ultra-powerful computer and that, by changing the computer software from time to time, he sends a series of stimuli to our brains so that we "experience" the same situations that we experience in real life. We could, for example, involve ourselves in a long and fulfilling love story, with its innumerable vicissitudes of joy, sadness, and languor, without ever entertaining the suspicion that any of this is unreal. It is a hypothesis that the skeptic uses to ask the provocative question, "How do you know that you are not a brain in a vat hopelessly lacking in knowledge concerning the reality in which you live?," but which at the same time exemplifies the reality configured by metaphysical realism. A case like the case of BIV can occur because between mind and the world there is an epistemic discrepancy, a gap, a separation that may not be eliminated at all, since for the metaphysical realist there is no metaphysical-epistemological guarantee in this sense.

We will return to the BIV hypothesis later, when we focus on the version of realism elaborated by Putnam in the 1980s, the one to which he gives the appellation "internal" and which he corroborates by presenting a radical critique of the hypothesis.

A Fifth Assumption: The OTT

There is a fifth assumption of the metaphysical realist that remained implicit in the above. In the event that human cognitive activity could fully unfold by successfully describing the entire world linguistically, the resulting global theory would be unique. Indeed, since the truth of any statement is explained on the basis of its "correspondence" to the actual situation it describes, and since the world possesses an intrinsic structure that is fixed and unambiguous, there can only be one global correspondence: the one that gives rise to the so-called *One True Theory* (OTT). The belief in the existence of a OTT—whether it can in fact be achieved or not—can therefore be regarded as another assumption typical of the metaphysical realist:[4] should he or she be confronted with two true theories that describe the world in a complete way, he or she would assert that these theories are merely "notational variants" of each other, variants that are compatible with each other in that they say the same things albeit in different terms.

Although early in his career Putnam was a staunch metaphysical realist (though not fully aware of the implications of such a position) and even presented an ingenious attempt to define (indirectly) correspondence (cf. Putnam 1960a)—leaving it without follow-up because he was little persuaded of its philosophical value—what set him apart from most other metaphysical realists was his conviction of the groundlessness of the belief in a OTT, something that, by his own admission, made him a *sophisticated* metaphysical realist (Putnam 1978a: 51 and 131). We have already encountered the reason: it is one of the constant features of his philosophy to believe in the existence of equivalent descriptions of the world that are mutually incompatible—that is, *not* mere notational variants—and this constitutes *ipso facto* a denial of the belief in a OTT. We can add here only that, in itself, such a belief appears quite naive. On the one hand, it certainly does not represent something typically realist, since even an idealist could possibly hold it; on the other hand, probably no philosopher with sufficient familiarity with the results of contemporary logic would be willing to endorse it. The reason is simple: the OTT is a theory that, should we actually succeed in formulating it, necessarily satisfies most (or all) of the requirements that the evaluation standards we developed at the time of the formulation of the OTT impose on any theory; in particular, the OTT certainly possesses the characteristic of being *complete* (i.e., of containing, for every statement p of the language in which it is formulated, either p or its negation), for otherwise there will exist some statement q such that either it or its negation is true and that it does not belong to the theory, which therefore will not be unique. Now, Kurt Gödel's first incompleteness theorem suggests precisely that OTT cannot be complete. For it states that for any coherent formal theory that contains even an elementary part of arithmetic, it is possible to identify a statement (our q) that is true and that is not a theorem of the theory; and since the OTT will surely contain arithmetic, it follows that it—assuming it is formalizable—is necessarily incomplete. In brief, if this reasoning is correct, the OTT does not exist.

As Gödel showed, there is no point in putting statement q within the scope of the theory by adding it to the axioms: there is always the possibility of identifying another statement q' that is true and is not a consequence of the theory. Gödel's fundamental result thus leads not only to the assertion that there can be no OTT, but also to the denial that there can be a final limit of inquiry or, equivalently, that all problems can be solved.

This last point—connected to an idea of one of the fathers of American pragmatism, Charles Sanders Peirce—will prove useful in illustrating Putnam's internal realism.

INTERNAL REALISM

The image of the world to which metaphysical realism gives rise is that of a world sharply separated from all that is human—cognitive faculties, mental properties, language—a world so independent and nonhuman that a scenario such as BIV is possible, a scenario that illustrates how and why our cognitive attempts are doomed to inevitable frustration. As mentioned above, partly by virtue of a careful analysis of the arguments that Michael Dummett had been advancing since the late 1950s against realist metaphysics in general, and partly by virtue of his drawing closer to the positions of the classical pragmatists, Putnam began to move away from metaphysical realism by presenting a series of critiques in the name of a different and weaker kind of realism that seemed to him better to answer the "major problem" of philosophy—the problem of how language and thought "hook on" to the world.

This series of critiques unfurls in articles, books, and lectures, many of which can be found in Putnam 1977, 1978a, 1981, 1983a, 1987a, 1988, 1990a, 1992b, 1994a—albeit with diminishing conviction, as he was destined to abandon even this kind of realism of his, *internal realism*. Compared with the former, we have here a realism "with a human face," in the sense that answers to questions related to "what the world is like" and "what is truth" receive an important contribution from the human mind—from human knowledge, to be more precise.[5] To render this idea Putnam uses the following metaphor (which he would regret years later):

> the mind and the world jointly make up the mind and the world. (Or [. . .] the Universe makes up the Universe—with minds—collectively—playing a special role in the making up) (Putnam 1981: xi).

The first time the name of this new realist position of his appears in print is in Putnam 1977, where it is considered, oddly enough, an empirical theory. This oddity would be explained some thirty years later: it had been an "unfortunate slip," since Putnam actually meant to refer to scientific realism—a viewpoint never abandoned at any phase of his career—and not to internal realism (cf. Putnam 2012c: 53). And indeed, if we reflect on what we saw in Chapter 5 (and as stated at the beginning of Putnam 1977), only scientific realism can supply the viewpoint that helps to explain certain facts, just as empirical theories do. The facts in question are the tendency of scientific theories to "converge," in the sense that earlier theories are often limiting cases of later ones, and the contribution that the use of language makes to achieving the theoretical and practical aims of speakers. Scientific theories converge by explaining a certain group of phenomena with an increasingly inclusive and more precise apparatus—and by fitting into each other like the cylinders of a telescope—because they gradually constitute better representations of an independent reality (which allows theoretical terms to retain their reference by moving from one theory to another), while the use of statements makes us achieve our purposes, when it is successful, because those statements capture actually existing aspects of reality—that is, they are true. Picking up on an idea already formulated in Putnam 1960a, Putnam states that

> The realist explanation, in a nutshell, is not that language mirrors the world but that *speakers* mirror the world—i.e. their environment—in the sense of *constructing a symbolic representation of that environment* (Putnam 1977: 483).

Scientific realism is nothing but this scientific picture of the relation of speakers to their environment and the role of language, a picture that emphasizes its character as an empirical hypothesis, an integral part of the theoretical scaffolding of science in general: "*science*'s explanation of the success of science, rather than as a metaphysical explanation of the success of science" (Putnam 1992g: 353). The reason behind the "unfortunate slip" is that Putnam intended precisely to convey the idea that realism is "internal" to science (cf. Putnam 2012c: 56), which is the idea we encountered in Chapter 5 according to which realism is *science's philosophy of science*.

To better understand Putnam's new position, it is necessary to keep in mind that the shift from metaphysical realism to internal realism is a shift from a non-epistemic to an epistemic conception of the world and truth. What counts as a world and what counts as a true statement is determined in part by human cognitive faculties, and this significantly attenuates that characteristic of any realist position, which is the independence of the world. Why then does Putnam believe that he can nevertheless remain in the realist field, despite the fact that the dependence of the world and truth on the human mind and its cognitive faculties has traditionally been considered an idealist trait? The answer is simple, though controversial.

World and truth spring from the knowledge embedded in our theories—belonging to both the natural and human sciences—but not from the theories currently formulated. It is not from these that a faithful description of reality can be derived (as would be the case in an idealist position), because knowledge progresses incessantly and in unpredictable ways, and for all we know the best theories of the moment may be partly wrong. No doubt, various aspects of reality are described by theories and statements that are true, but the fact is that some of these truths can only be reached by continuing the search far ahead in time. This does not detract from the fact that internal realism has an epistemic character: world and truth depend on our knowledge but—Putnam asserts in a move reminiscent of a typical idea of Peirce's—not on actual knowledge but on knowledge attainable under *ideal* epistemic conditions. As far as we can know, from actual knowledge world and truth are mostly independent (an idea that saves the trait of independence typical of the realist). The two expressions just used, "mostly" and "as far as we can know," deserve a clarification.

All it takes is a minimum of reflection to realize that much of our current knowledge is indeed *genuine* knowledge, that is, not only well-tested but also true. This is surely the case with knowledge expressed by simple, uncontroversial statements: we all know whether we have had at least one coffee on the day we read these very lines, and we know that Italy is washed by the Mediterranean Sea. These and other less trivial examples show that we already possess a significant amount of knowledge. Equally significant, however, is the amount of "alleged" knowledge, that which is expressed by statements for the truth of which we have good but not definitive justification—as is the case with most statements and theories of high-level empirical science. For the most part, the knowledge we may currently have at our disposal does not help us determine whether these are actually true statements or not.

Moreover, the peculiar human condition is such that even in the event that we were able to formulate a true statement, we may not realize it. For all we know, that statement is true, but we cannot rule out the possibility that there is some error in the process that led us to believe it to be true. This attitude, once brought to epistemological awareness, is nothing more than an expression of *fallibilism*, the thesis that—given our inherent fallibility—it is legitimate to doubt our cognitive acquisitions, provided we have a plausible reason for doing so. It is a typically pragmatist thesis that as we know Putnam always subscribed to but which now receives special emphasis in internal realism—"I should have called it pragmatic realism," he says at one point (Putnam 1987a: 17), "although the word 'pragmatism' has always been so ill-understood that one despairs of rescuing the term" (Putnam 1983c: 225).

Truth and Assertibility

The reason why it would not have been misleading to use the label "pragmatic realism" emerges perhaps best from the following consideration. Although the use of the adjective "internal" finds its justification in the fact that "to hold that *what objects does the world consists of?* is a question that it only makes sense to ask *within* a theory or description" (Putnam 1981: 49), which in some respects brings Putnam closer to the Quine advocate of *ontological relativity* (cf. Quine 1968),[6] the fact is that we adopt a theory or description to the extent that they prove useful in helping us to achieve our cognitive and practical purposes, aiding us in the formulation of an adequate representation of reality. Far from being able to avail itself of an absolute point of view such as that of the God's Eye point of view, all that our cognitive activity is substantiated by are "the various points of view of actual persons reflecting various interests and purposes that their descriptions and theories subserve" (Putnam 1981: 50). And this has a distinctly pragmatic flavor.

It is in the context of this realist option that the already-encountered phenomenon of equivalent descriptions and conceptual relativity finds its central place. At bottom, this option amounts to a more pronounced claim that "realism is *not* incompatible with conceptual relativity," and to considering "our familiar commonsense scheme, as well as our scientific and artistic and other schemes, at face value, without helping [oneself] to the notion of the thing 'in itself'" (Putnam 1987a: 17). This allows us to say that, depending on our interests and purposes,

> there *are* tables and chairs and ice cubes. There are also electrons and space-time regions and prime numbers and people who are a menace to world peace and moments of beauty and transcendence and many other things (Putnam 1987a: 16).

It all depends on which conceptual scheme we decide to use, relating our statements to it and asserting this or that statement as long as it is consistent with other statements within the scheme and supported by adequate evidence (empirical or otherwise), which evidence is capable of constituting sufficient epistemic guarantee for its assertion and which is indispensable for a truth claim to be raised with that assertion.

Thus, in the perspective of internal realism, the concept of truth is found to have very close ties with the concept of assertibility. To be precise, with that of *warranted assertibility* (a concept used by several classical pragmatists): it is epistemologically legitimate to assert a sentence *p*—or, equivalently, to rationally accept it—if, and only if, one has an epistemic guarantee for the assertion of *p*. Caution, however: this according to Putnam does not authorize one to consider the sentence true. If we were to identify truth with *current* warranted assertibility—that is, with what is justified by the knowledge inferable within the conceptual scheme to which *p* belongs—we would have as many truths about the same phenomena as there are conceptual schemes within which it is possible to justify *p*. Truth would then turn out to be "relative," contradicting one of our fundamental intuitions about a concept so central to our rational activity. Above all,

> truth cannot simply *be* rational acceptability for one fundamental reason; truth is supposed to be a property of a statement that cannot be lost, whereas justification can be lost. The statement "The earth is flat" was, very likely, rationally acceptable 3,000 years ago; but it is not rationally acceptable today. Yet it would be wrong to say that "The earth is flat" was *true* 3,000 years ago; for that would mean that the earth has changed its shape (Putnam 1981: 55).

As we said above, the typically realist character of independence is saved here by considering truth (and the world) as independent of *current*

knowledge—that enclosed in the well-corroborated conceptual schemes we
have at present—though not independent of *any* knowledge. Truth depends
on human cognitive faculties, but—the internal realist Putnam argues by ide-
alizing the power of those faculties—on those faculties if they operate under
ideal epistemic conditions: it is on the knowledge that those faculties can
succeed in acquiring under those conditions that truth depends. Therefore, to
say that a statement *p* is true is equivalent to saying that it could be justified
should the epistemic conditions be ideal: "truth is an *idealization* of rational
acceptability" (Putnam 1981: 55) or, equivalently, an *idealization* of warranted assertibility.

The appeal to an "ideal" dimension has puzzled more than one commen-
tator. What value can it have, one wondered, to put the greatest weight of
a philosophical position on a viewpoint that by definition is unattainable?
Moreover, does such an appeal not make Putnam's position fall under that
of Peirce, known for theorizing that a belief can be said to be true or false
only at the ideal limit of inquiry? Moreover, if—by extending the idealization—we imagine the situation in which *every* statement of our language
has reached its ideal limit (even if they do not do so at the same time) and
bring all the statements together to form a single comprehensive theory, do
we not obtain, albeit ideally, that OTT whose possible existence Putnam
clearly denies?

Putnam has a good answer here, however. He points out that the ideal epistemic conditions of which he speaks are analogous to the "frictionless planes"
of which physicists speak: they are certainly not things that physicists can
obtain, yet employing them in their reasoning has some utility because they
can be approximated to a high degree. So can ideal epistemic conditions. If it
is the term "ideal" that is disturbing, it can be replaced with the term "good
enough,"[7] since we can leave it to the intuition of the individual speaker,
whether expert or not, whether the conditions for making a statement in an
epistemically correct way are good enough or not. Not only that: for a good
number of our statements, we already possess these conditions (consider,
for example, the statement we referred to above, "Today I drank at least one
coffee").

As for the overlap between Peirce and Putnam, we can say that there is
a very close similarity between them if we consider *individual* statements
and their epistemic situation, but no overlap or similarity if we consider
the totality of statements collected into a supposedly ideal overall theory
(assuming, it is worth pointing out, that this was ever Peirce's idea, which
scholars mostly tend to deny). It is beyond doubt that Putnam thought it
impossible to give *simultaneously* ideal conditions for the evaluation of all
truths—or for the solution of all problem. Indeed, he drew attention to the
fact that

[t]here are some statements which we can only verify by failing to verify other statements. This is so as a matter of logic (for example, if we verify "in the limit of inquiry" that *no one ever will verify or falsify p*, where *p* is any statement which has a truth-value, then we cannot decide the truth of *p* itself, even in "the limit of inquiry"), but there are more interesting ways in which quantum mechanics suggests that this is the case, such as the celebrated Case of Schrödinger's Cat. Thus, I do not by any means *ever* mean to use the notion of an 'ideal epistemic situation' in this fantastic (or utopian) Peircean sense (Putnam 1990a: viii).

In conclusion, internal realism deserves the name "realism" insofar as

the world was allowed to determine whether I actually am in a sufficiently good epistemic situation or whether I only seem to myself to be in one (Putnam 1994e: 18).

This indicates that, within the plurality of conceptual schemes available to us, it is possible to derive a kind of objectivity—*objectivity for us* (cf. Putnam 1981: 55)—that provides a normative level aimed at indicating how there is a vital distinction between "being right" and "thinking we are right" (cf. Putnam 1981: 122, and 1983c: 225), and thus *better and worse* epistemic situations *with respect to particular statements* (Putnam 1990a: viii).

Showing that he keeps to two constraints that throughout his entire career he has never abandoned—"(1) *Rightness is not subjective* [and] (2) *Rightness goes beyond justification*" (Putnam 1989c: 114)—Putnam makes internal realism succeed in distinguishing itself as much from metaphysical realism as from relativism *tout court*: from the former because internal realism is epistemic in nature, and from the latter because it guarantees adequate space for a genuine notion of objectivity.

But why, as we mentioned above, does all this bring the phenomenon of conceptual relativity front and center?

The Critique of Correspondence: Absolute Skolemization

The reason is that internal realism constitutes a denial of the five assumptions listed at the beginning of the chapter, and in particular of the fourth one—the one concerning the existence of a fixed correspondence relation binding language (and mind) to the world. Putnam argues against correspondence by exploiting some theorems of logical model theory, then pointing out that conceptual relativity is something that cannot be dispensed with: the idea of correspondence is a metaphysical fantasy, he argues, while conceptual relativity is a fact.

To put it briefly, the problem Putnam sees with correspondence is that there are many ways of putting the symbols of a language and the things of a set S in correspondence with each other: "in fact infinitely many if the set S is infinite (and a very large finite number if S is a large finite set)" (Putnam, 1983c: 206). On what basis does Putnam substantiate this claim?

He makes interesting use of the so-called "Skolem-Löwenheim paradox," with the aim of extending the conclusions that Thoralf Skolem himself drew from it. The paradox is a consequence of the well-known theorem related to the two logicians' names, according to which every first-order theory that has a model has a model with a denumerable domain. Since a model is an interpretation that assigns to the terms of a theory its own set of objects making all statements true, the existence of multiple models for the same theory multiplies the number of objects to be matched by each term, defeating the idea of correspondence to a single world. According to Putnam there are two philosophical positions that can take advantage of the paradox: a verificationist position and a Platonism that "posits non-natural mental powers of directly 'grasping' forms" (Putnam 1980: 1). The implausibility of this positing, however, is for Putnam sufficient reason to declare verificationism the only alternative in the field. Let us see then first of all how the paradox arises.

Skolem posed the problem of whether set theory succeeds in uniquely describing the world of objects it deals with—sets. Putnam invites us to consider the following statement:

(i) $\sim(\exists R)$ (R is *one-to-one* $\wedge$ the domain of $R \subset N \wedge$ the range of values of R is S)

where N is the symbol denoting the set of all whole numbers and R is a binary predicate variable. If we replace S with the symbol denoting the set of all real numbers within a formalization of set theory, say that of Zermelo-Fraenkel, we obtain a theorem (Cantor's theorem, according to which the real numbers are non-denumerable). Because of this theorem, set theory states that the set S (understood as the set of real numbers) is non-denumerable, and therefore admits only non-denumerable models. But the Skolem-Löwenheim theorem tells us that no theory can have only non-denumerable models, thus giving rise to the paradox in question.

As is well known, the paradox is only apparent and does not constitute a real antinomy. In fact, its solution lies in the fact that set theory—of which statement (i) is a theorem—asserts that the set S is non-denumerable when the existential quantifier in (i) is interpreted as a quantifier that varies over all relations defined between N and S, that is, among all subsets of the Cartesian product $N \times S$. However, if one interprets the theory according to a model with a numerable domain, that quantifier does not vary over *all* relations, but only over relations *in the model*. In the numerable model, statement (i) is still true, and it states that there is no relation in the model that puts the set S in one-to-one correspondence with a subset of the whole numbers. The set S is thus non-denumerable, but it is so in a *relative* sense: in fact there are one-to-one correspondences between S and N, but these lie outside the numerable model. This shows, Putnam concludes, that

> what is a "countable" set from the point of view of one model may be an uncountable set from the point of view of another model. As Skolem sums it up, "even the notions 'finite,' 'infinite,' 'simply infinite sequence,' and so forth turn out to be merely relative within axiomatic set theory" (Putnam 1980: 2).

The lack of a model that fits set theory better than any other—that is, the lack of an *intended* interpretation of the theory, the one that the theoretical constraints and the operational constraints[8] of the theory seem to require—makes it impossible to capture the "intuitive" notion of a set, thus giving rise to relativity (cf. Skolem 1967).

Now, Putnam believes it is possible to extend Skolem's conclusions not only to the language of science, but also to the entire language. In the first case, that of the language of science, he advances the hypothesis that set theory could be used to formalize the conceptual apparatus of science as a whole. One would then have that

> even a *formalization of total science* (if one could construct such a thing), or even a *formalization of all our beliefs* (whether they count as "science" or not), could not rule out [. . .] *unintended* interpretations of this notion [of set] (Putnam 1980: 3),

and this shows how theoretical constraints would be of no help here. At this point, however, even if the theoretical constraints of the total scientific theory are knocked out by the theorem, one might think that the operational constraints succeed where the theoretical ones fail. In other words, one might believe that careful consideration of the relationships between theory and experience could identify *one* interpretation as the only one capable of preserving those relationships. However, Putnam rules out this possibility too.

Indeed, he draws attention to the existence of a "strong" version of the Skolem-Löwenheim theorem—the so-called "downward Skolem-Löwenheim theorem"—according to which a first-order theory that has a model, has a countable model that is a submodel of any given model.[9] Even if we allowed the existence of a denumerable infinity of measurable "magnitudes," and allowed the possibility of measuring each magnitude accurately correctly explaining and predicting phenomena, we would still find a countable submodel of the *standard* model that makes the same statements true by

preserving the way the measurement procedures of those quantities are interpreted. Even operational constraints, then, can be satisfied by both intended
and unintended interpretations. Therefore, one cannot help but conclude that
the whole language of science, not just the portion used for the formulation
of set theory, is vulnerable in the face of the theorem.

But, as we have mentioned, the extension of Skolem's conclusions can be
taken even further, coming to affect our entire language: not only the portion
which is used to express mathematics and science, but also that which we use
to report ordinary experience.

It follows that every word we use in everyday speech, every term employed
in casual conversation, whether it is about concrete empirical objects or
abstract unobservable entities, every fragment of language may have a plurality of different references. Our entire language will be nothing more than a
formal construct interpreted differently depending on the model under consideration from time to time. We could comfortably talk about our cat resting on the mat in front of the fireplace, and one of our interlocutors—solely
because he or she interprets the words of the very same language according to
another model—might think we are talking about the cherries hanging from
the tree in the garden outside.[10]

> In short, one can "Skolemize" absolutely everything. It seems to be absolutely
> impossible to fix a determinate reference (without appeal to non-natural mental
> powers) for *any* term at all. And if we apply the argument to the very metalan-
> guage we use to talk about the predicament...? (Putnam 1980: 16)

It is thus here—to absolute Skolemization—that Putnam's analysis leads.
If Skolem had thought that he could conclude, in the face of the paradox that
bears his and Löwenheim's name, that every notion of set theory is relative
and that therefore this theory is unable to describe unambiguously its field
of inquiry, Putnam's extension of this conclusion goes so far as to affirm the
relativity of even the reference of the most common terms, where this relativity is understood as dependence on the theory to which the term belongs. It
is the world itself in its totality that does not constitute a unique and uniquely
describable set of objects, a set that stands firmly in front of every possible
theory determining a correspondence relation. This typical metaphysical realist image crumbles, giving way to what according to Putnam is a more plausible image, closer to our condition as thinking and speaking beings—more
human. It is an image that highlights how

> signs do not intrinsically correspond to objects, independently of how those
> signs are employed and by whom. But a sign that is actually employed in a

particular way by a particular community of users can correspond to particular objects *within the conceptual scheme of those users*. "Objects" do not exist independently of conceptual schemes. *We* cut up the world into objects when we introduce one or another scheme of description. Since the objects *and* the signs are alike *internal* to the scheme of description, it is possible to say what matches what (Putnam 1981: 52).

As highlighted by the phenomenon of conceptual relativity,

> "objects" themselves are as much made as discovered, as much products of our conceptual invention as of the "objective" factor in experience, the factor independent of our will (Putnam 1981: 54),

without its being possible to determine which is the conventional and which the factual part of our language. To claim to do this would be to commit a *fallacy of division* (cf. Putnam 1990a: x), and implicitly to adopt that cookie-cutter metaphor that in Chapter 2 we saw Putnam flatly rejects.

This is why the phenomenon of conceptual relativity assumes central importance in internal realism: it is here that this phenomenon finds its proper place, in spite of what the sophisticated metaphysical realist Putnam believed. Since the supposed correspondence relation is something "external" to both the mind and the world, to describe it would require a superhuman point of view that can cognitively reach where human beings cannot. If, therefore, we want to provide a picture of our metaphysical situation that is at least adequate, the idea of conceptual relativity imposes itself on us. The important thing to bear in mind however, Putnam warns us, is that "urging this relativism is not advocating *unbridled* relativism" (Putnam 1980: 10): objective (if evolving) canons of rationality *do* exist, although they are unable to settle something like the issue of correspondence to the metaphysical realist's satisfaction.

Thus, only *within* conceptual schemes can a correspondence relation be traced, not independently of them. In addition to leading to a different interpretation of the notion of correspondence from that of the metaphysical realist, this also leads to a thesis that the internal realist Putnam held with great conviction: the thesis that *if there were an ideal theory, it could only be true*—contrary to what a metaphysical realist would say.

An ideal theory would be the best theory about the world that we could ever hope to formulate, the theory that possesses all the possible requirements that we believe any product of our cognitive faculties should have.[11] An understanding of "ideal theory" that might fit Putnam's thesis can be conceived by a being endowed with superhuman intellectual powers, a "being with a different spatiotemporal perspective [from ours], or whose observational and

intellectual powers transcend our own, such powers being modelled on those
which we possess, but extended by analogy," as Dummett puts it (Dummett
1976: 60).[12] Putnam's discussion of the ideal theory employs the notions of
theoretical constraint and *operational constraint*. We have already encountered them (see footnote 8) and noted that although considered by logical
positivists as constraints on the acceptance of theories, they can also be taken
as constraints on their interpretation.

The operational constraints that a theory must satisfy in order to be accepted
prescribe the executability of certain operations if it follows from the theory
that those operations must be performed. If, say, the statement "Electricity
flows through this wire"—let's call it *E*—belongs to the theory, then the
needle of the voltmeter we prepare to verify the phenomenon *must* move (at
least in most cases, considering possible disturbing effects). The operational
constraint can be in this case something like, "The theory is acceptable if and
only if, most of the time, statement *E* is true when experimental condition *C* is
satisfied," where condition *C* is the control contrived to appraise the theory's
performance. Therefore, it can be said that operational constraints "connect
the theory to the world," trying to test its experimental adequacy: they attempt
to establish a correspondence between what the theory says and experience.

Theoretical constraints, on the other hand, refer to the formal properties
that a theory must have in order to be accepted. They require it to possess a
certain number of axioms, to be elegant, consistent, plausible, functionally
simple,[13] conservative, and so on. The latter constraint, for example, might
state something like, "Do not accept a theory that requires the elimination of
a large number of previously accepted beliefs if another equally simple theory
is available that preserves those beliefs and agrees with the observations" (cf.
Putnam 1981: 31).

Both constraints, then, are an aid to the acceptance of theories. However,
they can play the same role with regard to the interpretations that can be made
of those theories. Between "acceptance" and "interpretation" there is a close
connection. We never accept a theory that does not already, implicitly, have
a certain interpretation. A theory understood simply as a formal structure
merely provides a certain vocabulary to be applied to the world, but still says
nothing about *how* this world is, in what way that vocabulary is to be applied
to it, what kinds of relations and properties are in question. An interpretation
does just that: it specifies how and how big the world is to be, establishing a
set of properties and relations among its objects. However, not every interpretation is as good as any other. It is therefore necessary to narrow down
the class of permissible interpretations of a theory, and both operational and
theoretical constraints can be used for this purpose.

Now, on the basis of these two constraints Putnam invites us to consider
an ideal theory, T_I, that

can be imagined complete, consistent, to predict correctly all observation sentences (as far as we can tell), to meet whatever "operational constraint" there are [. . .], to be "beautiful," "simple," "plausible," etc. (Putnam 1977: 485).

Putnam also invites us to assume that the world contains an infinite number of objects, and T_I asserts that there is an infinite number of things. T_I will then have only models with an infinite domain. Since T_I is consistent, it follows from some results of model theory that it has a model of every infinite cardinality.[14] We then choose a model M of the same cardinality as the world and apply the elements of the domain of M one by one to the objects of the world, such that we define the relations of M directly in the world (modifying the interpretation of predicate letters accordingly). We thus obtain a model M' that has the same world as its domain and that specifies a satisfaction relation SAT between the terms of the language of T_I and the objects of the world, where "TRUE" will be the truth predicate, defined *à la* Tarski, determined by relation SAT. T_I is then true of the world, as long as "true" is interpreted as TRUE.

Now, Putnam says, the metaphysical realist—wanting to deny that the ideal theory T_I is necessarily true—could argue that the SAT relation is not the "intended" correspondence between T_I's language and the world. Which in turn implies that it is not the one that meets the operational and theoretical constraints that the superhuman being mentioned above would impose on T_I. But—and herein lies the core of Putnam's argument—T_I's ideality consists precisely in satisfying both constraints. If, say, T_I did not satisfy all the operational constraints, it would contain at least one false statement and would be discarded. This, however, does not happen with T_I. Let us take, for example, a true statement, K. Since its negation is false, and thus cannot belong to T_I, and since T_I is complete, K is a theorem of T_I. From this and from the fact that SAT—by construction—is a true interpretation of T_I, it follows that K is TRUE (in the sense determined by SAT). On the other hand, suppose that the statement K is false. It is therefore not a theorem of T_I, but its negation will be—because of the completeness of T_I. This then will be TRUE, since, as we have seen, T_I is TRUE (in the sense determined by SAT). From this and from the fact that T_I is consistent, it follows that K is FALSE.

It becomes clear at this point that the interpretation M', which specifies SAT as a correspondence relation between the language of T_I and the world, allows the ideal theory to satisfy every possible operational and theoretical constraint that would be imposed at the ideal limit of inquiry. So, Putnam concludes,

what *further* constraints on reference are there that could single out some other interpretation as (uniquely) "intended," and SAT as an "unintended" interpretation (in the model-theoretic sense of "interpretation")? The supposition that even an "ideal" theory (from a pragmatic point of view) might *really* be false appears to collapse into *unintelligibility* (Putnam 1977: 486).

Can we therefore consider the metaphysical realist defeated by this argument of Putnam's? Is the ideal theory a neutral confrontation ground on which the metaphysical realist is inevitably defeated? Some considerations lead us to give a negative answer to these questions.[15]

Some Considerations

First of all, it is difficult to think that a metaphysical realist would so easily accept the notion of an ideal theory. After all—she might argue—the very notion of "ideality" is *our own* notion; it is *we* who determine how and where to remain satisfied with the cognitive qualities of a certain theoretical production. And even if we did this entirely "objectively"—that is, even if what we isolated as an ideal theory were *truly* so, given the power of our cognitive faculties—the fact remains that the world might contain aspects at odds with the theory we have thus isolated. For the world, the metaphysical realist would continue, is independent of our theories, including the ideal theory. What is ideal according to us might not be ideal "according to the world," so to speak. An ideal theory might be appreciated only by God, only by someone who is able to survey the world in a single glance.

Putnam might retort that his argument moved on essentially the same iron lines on which logicians operate with their tools, and that therefore it is hardly possible to escape certain conclusions unless one rejects much of what logicians have achieved in this century, and thus unless one rejects logic itself. But the metaphysical realist would point out to him again that it is *we* who set the theoretical and operational constraints by means of which—according to Putnam—interpretations of a theory may be creamed off. It is for us that a theory counts as simple or elegant, and it is we who—however idealized—establish a set of relations between theory and experience. Therefore, who can say that the constraints we might set on the ideal limit of inquiry are those that the world sets—or, rather, has already set? The world, after all, can transcend our ability to provide interpretations.

We can only conclude, at this point, that the ideal theory—whichever way you look at it—will never be accepted by a metaphysical realist, who for that very reason will not feel touched at all by the argument Putnam sets up. Indeed, the idea of the existence of an ideal theory can only make sense to a philosopher with an epistemic orientation. As Dummett says, "this line of thought is related to a [. . .] regulative principle governing the notion of truth: if a statement is true, it must be in principle possible to know that it is true" (Dummett 1993: 61). Putnam himself seems to admit this—albeit unconsciously and, literally, parenthetically—in the above quotation: indeed, he states that the supposition of the falsity of the ideal theory collapses into unintelligibility when made from a pragmatic point of view. But it is precisely this point of view that is unacceptable to a metaphysical realist, for here "pragmatic" is synonymous with "verificationist," and thus with "epistemic." Therefore, the notion of an ideal theory is not metaphysically neutral—it does not constitute a neutral confrontation ground between realists of different kinds. Moreover, in thinking that his is a metaphysically neutral argument—that is, an argument you can accept regardless of your metaphysical inclinations, an argument that you should accept, say, as you accept a correct mathematical proof—Putnam indirectly shows himself to be too much in the grip of internal realism, so much so that he ends up by surreptitiously assuming it. And here a *circularity* comes to light.

The problem, simply put, is this. According to the results of Skolem and Löwenheim and the extension made of them by Putnam, there is an infinity of models for the ideal theory, that is, an infinity of interpretations according to which statements of the theory are true. The problem of finding a model to be considered "intended," then, a model that specifies the only right reference relation, should be a problem for Putnam as well, since his own use of model theory leads to the equal validity of so many such relations. How is it, nevertheless, that Putnam remains satisfied with having identified SAT? How come he declares that the correspondence relation specified by SAT is the good one?

In response it will be convenient to establish what kind of philosopher can consider such a result "plausible." We have seen that the problem arises not from the fact that there is no referent for every word we employ in our speech, but from the fact that there is an infinity of them. One does not deny that we can have a referential use of language: one denies that such use is unambiguous, and that we can therefore succeed in agreeing on the referent of our terms. In short, it is denied that—through proper cognitive procedures—it is possible to assign a definite referent to each term. What is banned therefore is the *knowability* of the referent relation. But then, if this is the case, the proponent of such a position can only be a *radically non-epistemic realist*, that is, a realist who strongly denies any possibility of knowledge—if there ever was such a philosopher (what in the literature might be called a "radical skeptic"). It is the radically non-epistemic realist who can positively say, "We are hopelessly subject to absolute Skolemization." (Just as the proponent of the BIV hypothesis says, "We all are and always have been BIV," as we shall see below.)

If this is the case, we have at least two positions that are immune from the danger of absolute Skolemization: a *simple* (or *moderate*) *non-epistemic realism* and an *epistemic realism*, since both admit, albeit to different degrees, the possibility of knowledge. These are labels we have not used so far, and since they can help us in orienting ourselves within the family of realist positions, it is worth elucidating them briefly.

The *simple non-epistemic* realist admits the possibility of knowledge of the world: what she denies is that we human beings can have any "guarantee" that we can achieve such knowledge at any given time. According to her, we cannot rule out the possibility that some parts of the world will remain unknown to us, even in principle. Nevertheless, she considers our cognitive capacities to be suitable for obtaining knowledge of the world, and therefore her position has no skeptical outcomes—like the *radically non-epistemic* realist, for whom the power of human cognitive faculties is so limited that they cannot reach the world. The *epistemic* realist on the other hand (of which the internal realist Putnam is an example) holds that we possess guarantee of cognitively reaching the world, even if under certain conditions, and that therefore the world has no aspects that lie beyond the power of our cognitive faculties. It is like saying that the whole of reality is knowable, even if only in principle.

Now, if—assuming the necessary countermeasures—even a certain kind of metaphysical realist (the simple non-epistemic one) can curb the Skolemization applied to ideal theory, it follows that Putnam's argument that ideal theory cannot be false—since a certain model ultimately turns out to be *intended*—is "weak," because its validity—assuming there is any—is limited solely vis-à-vis the radically non-epistemic realist. Unfortunately, however, the surreptitious assumption of internal realism noted above makes it impossible to consider the argument convincing.

Putnam in fact arrives at verificationism by assuming verificationist premises. He "assumes"—without saying so—that the problem encountered by the metaphysical realist does not arise from a verificationist standpoint, a standpoint suggested to him by Michael Dummett. In both Putnam (1977) and Putnam (1980) he explicitly states that the thorny question of correspondence *dissolves* once we attain what he judges to be the only right perspective. But such a perspective is something the metaphysical realist will always reject if she wants to respect her most fundamental beliefs. If she is a radically non-epistemic realist, she will use the Skolem-Löwenheim theorem to illustrate her peculiar skeptical beliefs about reference, and she will be quite satisfied by being able to point to a situation—absolute Skolemization—that faithfully reflects what she goes on to hold about our cognitive situation. If, on the other hand, she is a metaphysical realist who admits the possibility of knowledge, she too will be able to use the Skolem-Löwenheim theorem as she wishes and yet declare herself safe from any possible referential impasse.

That being said, note that the difficulty of giving an account of the correspondence between thoughts, or words, and supposedly mind-independent objects has been a difficulty well known to philosophers at least since the time of Kant. German idealism itself began its intellectual journey in the nineteenth century by highlighting the fact that this is an insurmountable difficulty. It would seem then that this is an established fact of Western philosophical inquiry, that there are no longer philosophers willing to recognize themselves in the coordinates of metaphysical realism, and that therefore Putnam is criticizing an artfully constructed opponent (apart from criticizing an earlier image of himself). In fact, as is often the case in philosophy, old ideas are posed again exploiting new conceptual tools—a procedure that is vital to philosophy itself and has sometimes led to new, more mature awareness about problems repeatedly addressed in the secular history of Western thought. This is precisely what the metaphysical realists of our time have attempted to do. They are therefore flesh and blood human beings—by no means artfully constructed.

The Causal Relationship

In our age, Putnam argues, the program of metaphysical realism has been taken up by those who subscribe, in one form or another, to *materialism*: "Today material objects are taken to be paradigm mind-independent entities, and the 'correspondence' is taken to be some sort of causal relation" (Putnam 1983c: 205). If by using a certain term I am able to refer to an object, a materialist would say, this is because I have had a number of causal transactions with that object in the past, where the causal relation that allowed me to have such transactions is *inscribed* in the material structure of reality itself. Theoretical and operational constraints may not be able to isolate a single correspondence as that "intended," but—the materialist would continue—constraints capable of doing this do exist: they are those offered by nature itself. Being part of material reality, it is the task of the natural sciences to arrive at a description of the causal relation, the reference relation and thus the correspondence. The program of materialism—of *physicalism*—would therefore permit the formulation of a "natural" metaphysics, guided by the scientific method and falling within the reassuring confines of the natural sciences.[16]

The question then arises whether a causal relation that is "inscribed" in the material structure of the world really exists, and whether it can be used to explain the semantic reference relation between language and the world: in short, "is *causation* a physical relation?" (Putnam 1983c: 211) A simple and immediate answer is that "Nature, or 'physical reality' in the post-Newtonian understanding of the physical, has no semantic preferences" (Putnam 1984a: 83), while a more circumstantial answer cannot but start by noting how the idea of physical causation implying a real and *necessary* relation clashes with the serious doubts that David Hume's analysis introduced into philosophical thought more than two centuries ago.

By distinguishing sharply between what "exists in reality" and what is 'projected' by the human mind into reality, Hume argued that the causal relationship depends only on a psychological *habit* of expecting the occurrence of a certain type of event immediately after the occurrence of another type—for example, expecting thunder after seeing lightning—without there being the slightest necessary connection between the two types of events. To believe that there is a causal link between the two types is merely our own projection. Conversely, then, what the physicalist does is to assume the existence of "non-Humean," and therefore *physically real* causation. Two paths are open to him: the first is to show how such a supposed physical relation is *definable* on the basis of fundamental physical quantities (field tensors, etc.).

This is tantamount to involving oneself in an enormously complex reductionist program that is difficult, if not impossible, to implement, but let us assume that the way is viable. A cause would then be a sufficient condition for its effect: as soon as it happens, the effect must follow (at least in a deterministic world). Harkening back to John Stuart Mill, Putnam points out that in ordinary language "cause" hardly ever means such a cause—a *total cause* as Mill calls it.[17] The total cause of a forest fire consists of a multiplicity of elements (the dryness of the leaves, their proximity to a campfire, the temperature of a given day, the presence of oxygen in the atmosphere, and so on): only some of these elements are chosen when we have to answer the question "What caused the forest fire?" And we choose this or that element according to its usefulness given our purposes and interests in providing an explanation.[18] This shows, on the one hand, that in its ordinary sense "to cause" can very often, if not always, be paraphrased using a phrase involving "to explain," and, on the other, that we consider certain parts of the total cause as *background* by selecting only the part that interests us as *the* cause. And all this "is not written into the physical system itself" (Putnam 1984a: 87): what is or is not a cause, or an explanation, depends on background knowledge and our reasons for giving a certain answer. From this it follows that

> even if the notion of "total cause" *were* physically definable, it would not be possible to *use* it either in daily life or in philosophy; the notion the materialist really uses when he employs "causal chain," etc., in his philosophical explications is the intuitive notion of *explanation*. But this notion is certainly not physically definable (Putnam 1983c: 213).

The other path left for the materialist is to recognize that causation is not definable in the terms of the fundamental parameters of physics and to assume it as a "primitive" relation, arguing that it is science itself that needs to postulate it (this is the path in Boyd 1980). However, Putnam notes, "can a philosopher who accepts the existence of an irreducible phenomenon of *causation* call himself a materialist?" (Putnam 1983c: 215) As just seen, "causes" is paraphrasable with "explains," and so-called *causal powers* are "properties that *explain* something, given background conditions and given standards of salience and relevance. [But] salience and relevance are attributes of thought and reasoning, not of nature" (Putnam 1983c: 215). To hold that science requires such a thing not only leads to unduly inserting into science a metaphysical interpretation that physicists themselves have for centuries rejected as unserviceable for their purposes,[19] since "even the time-directedness of causal processes disappears in fundamental physics" (Putnam 1984a: 86),[20] but on closer inspection amounts to holding that causation "has a special intentionality (perhaps this is what its 'non-Humean' character consists in). If A causes B then A *explains* B; and explanation is connected with reason itself" (Putnam 1984a: 85). In other words,

> if events *intrinsically* explain other events, if there are saliencies, relevancies, standards of what are "normal" conditions, and so on, built into the world itself independently of minds, then the world is in many ways *like* a mind, or infused with something very much like reason (Putnam 1983c: 215–216).

Thus, in his stubborn attempt to regard the world as endowed with its own intrinsic structure that offers mind-independent objects linked by a predetermined causal relation, the materialist ends up with a world that does not at all meet the requirements he prejudicially wanted to see in it. Far from justifying the belief in the existence of an unambiguous and well-determined correspondence relation between language and the world—a relation that can be used to substantiate, among other things, the reference relation connecting words to objects—causality turns out to be a *cognitive* notion: an elastic notion that allows one to distinguish variously between cause and background condition depending on the epistemic interests of the moment. Likewise,

> reference, like causality, is a flexible, interest-relative notion: what we count as *referring* to something depends on background knowledge and our willingness to be charitable in interpretation. To read a relation so deeply human and so pervasively intentional into the world and to call the resulting metaphysical picture satisfactory (never mind whether or not it is "materialist") is absurd (Putnam 1983c: 225).[21]

But, one wonders, if the attempt to see a non-Humean cause-and-effect relation in the world is doomed to failure, and if such a relation is cognitive in character, does this mean that we have no choice but to embrace

Hume's position and conclude that both causality and reference are projections of the human mind? The answer reveals a central aspect of Putnam's philosophical position, an aspect that came to the fore in the 1980s through an increasingly acute sensitivity to issues of pragmatism and was never abandoned.

Indeed, he asserts that it is the very *dichotomy* between "what really exists" and "what we project" that is devoid of content: it is exactly the kind of distinction that requires the God's Eye View and thus the perspective of metaphysical realism—a perspective that has turned out to be absurd. Instead of asking whether causation exists "in reality" or not, what we should do is to take note that it undoubtedly exists in the *lifeworld*, as Husserl put it, which is real in a *phenomenological* sense. This is a lesson that can be drawn not only from Husserl, but also from Wittgenstein (cf. Wittgenstein 1976), Austin, Dewey, and James:

> the world of "ordinary language" (the world in which we actually live) is full of causes and effects. It is only when we insist that the world of ordinary language (or the *Lebenswelt*) is defective (an ontological "jungle," vague, gappy, and so on) and look for a "true world" (free of vagueness, of gaps, of any element that can be regarded as a "human projection") that we end up feeling forced to choose between the picture of "a physical universe with a built-in structure" and "a physical universe with a structure imposed by the mind" (Putnam 1984a: 89).

This is not to abjure materialism and try to formulate a non-materialist metaphysics:

> We cannot really go back to the Middle Ages or to Plato's time. If science does not tell us what is "really there" in the metaphysical sense, then neither does anything else. What has collapsed is the attempt to divide mundane reality, the reality of the *Lebenswelt*, into Real Reality and Projection (Putnam 1984a: 90).

The attempt to obtain a simplistic image of reality as a rigorously preordained set of objects and properties, with a battery of dichotomies playing an ordering role, has collapsed.

The Rejection of Dichotomies

Principal among these dichotomies is the one that sharply distinguishes between *objective* and *subjective*. Putnam's internal realist period itself was inaugurated with the aim of

> break[ing] the strangle hold which a number of dichotomies appear to have on the thinking of both philosophers and laymen. Chief among them is the

dichotomy between objective and subjective views of truth and reason (Putnam 1981: ix).

The examples of such dichotomies are well known (some we have already encountered): analytic/synthetic, observational/theoretical, fact/value, internal/external, actually existing/mere projection, intrinsic (or essential)/extrinsic properties, world-in-itself/concepts we use to think and talk about it, true/assertible, truth conditions/assertibility conditions (cf. for example Putnam 1991: 265), and the like. To hold that these dichotomies have a "suffocating" influence on our thinking means that they prevent us from getting a picture of the world, and of ourselves as part of the world, that is true to our real situation; it means that they confuse rather than help clarify, and that they do this because they are completely arbitrary—that is, lacking justification. It should be borne well in mind, however, that an awareness of the impossibility of making a clear distinction between the poles of these dichotomies in general, and that between the objective and the subjective in particular, is not the same as asserting that it is all subjective. "Rejecting the dichotomy *within* kinds of 'truth,'" for example, "is not the same thing as saying 'anything goes,'" says Putnam (1987a: 32) echoing Paul Feyerabend's well-known dictum. As we have already mentioned, internal realism intends precisely to be an alternative "to metaphysical realist views of reality and truth, on the one hand, and to cultural relativist ones, on the other" (Putnam 1987a: 1). What Putnam wants to emphasize is that we are not obliged to choose either pole of a dichotomy, because it is the dichotomy itself that is meaningless, just as the various aspects of the metaphysical realist framework are meaningless, or absurd, or self-refuting.

Of the latter type—self-refuting—is for Putnam the above-mentioned BIV hypothesis.

BIV: A Self-Refuting Hypothesis

In the course of his reasoning about the consequences of the Skolem-Löwenheim theorem and some of the considerations of the second Wittgenstein, Putnam comes to develop an argument on the basis of which he shows the self-refuting nature of the BIV hypothesis, an argument that was to initiate a proliferation of essays and articles—both favorable and of severe criticism—that persists at the time of the publication of the present volume (even a quick Internet search can give an idea of the debate that has developed on this subject from 1981 to the present).[22]

A self-refuting hypothesis is a hypothesis whose truth implies its own falsity—for example, "All general statements are false." Or it is a hypothesis that refutes itself as soon as someone utters it—for example, "I do not exist."

Of the latter kind is the self-refuting nature of the BIV hypothesis. The singular—and in many ways ingenious—route through which Putnam goes to prove such self-contradictoriness is based on the mechanism of reference that we have seen at the heart of the semantic conception of natural kind terms. In short, the idea is that although a BIV can apparently think or say whatever we can think or say, it cannot *refer* to what we refer to. In particular, it cannot think or say that it is a BIV (even using these same words). But why?

The reason is that a BIV lacks the *proper* causal interactions with objects in the world, those interactions that allow us to refer to cabbages when we use the word "cabbage." We are able to establish a proper causal relationship with objects in our environment, we are able to manipulate such objects by deriving a range of implicit and explicit information, which—as we saw in the chapter on Putnam's semantic conception—allows us to endow the words we use when we talk about these objects with a reference. As Putnam points out,

> the ability to refer to things is not something that is guaranteed by the very nature of the mind, as Descartes mistakenly believed; reference to things requires information-carrying interaction with those things (Putnam 2002a: 108).

It is true that BIVs possess an analogue of our nerve endings and sense organs through which they can receive input and transmit output (a transmission of information that occurs through the wires that connect their synapses to the super-sophisticated computer), but all they can come into causal contact with is what the computer makes them "see" when they think of a tree: it is the *image* of a tree, not an actual tree. It follows that all they can causally connect with and refer to using the word "tree" are

> trees in the image, or [. . .] the electronic impulses that cause tree experiences, or [. . .] the features of the program that are responsible for those electronic impulses (Putnam 1981: 14).

Therefore, when I—who, until proven otherwise, am not a BIV—think or say "I am a BIV," I use words that have genuine reference to brains and vats by virtue of my genuine direct and indirect interactions with brains and vats; but when a BIV thinks or says "I am a BIV," he uses terms that refer to, say, *images* of brains and vats, and therefore fails to articulate a thought or statement worthy of the name. Those causal interactions give *no information* about real brains and real vats. Putnam points out that his is a Kantian-style *transcendental argument*, since it highlights a number of "preconditions" of thought and reference. This is why from the absence of proper causal connection he concludes that a BIV cannot think or say that it is a BIV.[23] Based on this conclusion he then demonstrates the self-refuting nature of the hypothesis. Before following the proof step by step, let us see an abbreviated version of it. This is a version that Putnam used in his classes when explaining his argument to students.

The premises of the proof are two: one we have just seen, and that is the *causal constraint on reference*: one cannot refer to certain objects using a certain term unless one has had information-laden causal interactions with these objects (or with things or properties by which a description of these objects can be formulated). The second premise is the *disquotational property of reference*, according to which in my language any term *t* refers to (all and only) *t*'s. (Intuitively, this property is called "disquotational" because it results in a removal of quotation marks around a term, where the quoted expression is a name of a term, while the disquoted expression is a name of the object(s) to which that term refers.)

Finally, a definition—*vat English* is the language I would speak if I were a BIV—and here is the short version of the proof that I am not a BIV:

(1) In my language, "cabbage" refers to cabbages   [second premise]
(2) In vat English, "cabbage" does not refer to cabbages [first premise]

Therefore:

(3) My language is not vat English   [definition of vat English]

Therefore:

(4) I am not a BIV.

The short version[24] provides a clear idea of the move contrived by Putnam to refute the BIV hypothesis, but it does not illustrate how "stating" this hypothesis leads to its denial. This is more evident from the reconstruction of individual passages of the Putnam's argument that I provided in Dell'Utri (1990). Here it is (the premises are as above, "iff" stands for *if and only if*, and the equivalence thesis is the analogue of the disquotational property of reference applied, however, to statements):

(1) We are BIV [hypothesis]
(2) If we are BIV, then we are not BIV in the image [explication of the hypothesis]
(3) We are not BIV in the image [(1), (2)]
(4) If we are BIV, then every word we speak refers to objects in the image [first premise]
(5) Every word we speak refers to objects in the image [(1), (4)]

(6) Every statement *A* that we utter is true iff *A* is true in the image [(5)]

(7) The statement "We are BIV" is true iff it is true in the image [exemplification of (6)]

(8) The statement "We are BIV" is true in the image iff we are BIV in the image [equivalence thesis]

(9) The statement "We are BIV" is true iff we are BIV in the image [(7), (8)]

(10) The statement "We are BIV" is false [(3), (9)]

(11) The statement "We are BIV" is false iff the statement "We are not BIV" is true [definition of falsehood]

(12) The statement "We are not BIV" is true [(10), (11)]

(13) The statement "We are not BIV" is true iff we are not BIV [equivalence thesis]

(14) We are not BIV [(12), (13)]

Thus we have the conclusion that: if we are BIV, then we are not BIV (or—on a metalinguistic level—if "We are BIV" is a true statement, then it is false). "So it is (necessarily) false" (Putnam 1981: 15).

It should be remembered that this is an argument addressed by Putnam not only (and not so much) against a certain form of skepticism (the one Descartes attempted to respond to), but especially against metaphysical realism, since the hypothesis assumes its plausibility within a *radically* non-epistemic metaphysical framework: a framework within which even an ideal theory—assuming there is such a thing—could be false, despite the fact that it is our best cognitive attempt, as we saw above.

So, if our real condition as human beings is to be cognitively separated from the world, and irremediably so, so that we might remain forever in the dark about that condition, then we may be BIV without ever being able to know it. Conversely, if our condition is inscribed in an epistemic frame (such as that offered by internal realism), or if we are placed in a non-epistemic frame that nevertheless does not prevent the acquisition of genuine knowledge (such as that offered by a *simply* non-epistemic realism), then there is nothing to prevent us under sufficiently good epistemic conditions from being able to know whether we are BIV or not—at least in principle.[25]

Bear in mind, however, that Putnam presented his refutation of the BIV hypothesis while he was embracing internal realism, and some commentators saw such a close relationship with it that they considered the refutation fallacious—a kind of *petitio principii*, in that it seemed to tacitly assume the validity of internal realism (against this view, see Thorpe 2019 and Thorpe & Wright 2022). On the contrary, Putnam always considered his argument to be "objectively" valid, an argument that we should all therefore accept, including an intellectually honest metaphysical realist; and this—to return to the notion of cause—would stand to prove the falsity of the metaphysical realist's causal conception, since as we have seen *we* cannot be assumed to have available a notion of 'causation' that transcends *our* particular way of being situated in the world" (Putnam 1992f: 362–363).

The fact is that some people seemed unconvinced by Putnam's arguments:

> my critics [. . .] argued [. . .] that I had not *proved* that nature could not fix what our words refer to without regard to our intentions, and what I replied was that the idea of the world imposing reference on our terms or truth-conditions on our predicates makes no sense. But I did not consider [. . .] (Putnam 2012b: 78).

We will pick up on this "but" in Chapter 7, at the end of which we will try to give a final assessment of Putnam's argument against BIVs.

NOTES

1. The locution "ready-made world" is attributed by Putnam to Nelson Goodman, but Goodman borrows it from William James (see James 1907: 123).

2. The reference is to the saying attributed to Archimedes, "Give me a foothold (and a sufficiently long lever) and I will lift the world."

3. For an interesting discussion of the differences between Descartes's and Putnam's treatment of the skeptical hypothesis (as well as those of other contemporary thinkers), see Davies (2004).

4. Interestingly, Putnam attributes this assumption not only to the transcendental realism opposed by Kant, but also to Kant himself: cf. Putnam 1984b: 39–40.

5. Putnam himself has drawn attention to the analogy between his internal realism, which opposes the idea of a ready-made world, and Kant's transcendental idealism. On the scope and limits of this analogy, as well as the problems this Kantianism encounters, see Moran (2000).

6. It brings Putnam closer to Quine without, however, registering a coincidence. Indeed, Putnam tends to distance himself from Quine on this point as well: his argument based on model theory, which we shall meet below, for him "is not a proof of ontological relativity but rather a *reductio ad absurdum* of ontological relativity" (Putnam 1994f: 251). The relationship between his model-theoretic argument and Quine's is discussed in Putnam (1989b).

7. In this regard Putnam quotes, in several places and with great approval, John Austin's saying "enough is enough: it doesn't mean everything" (Austin 1946: 84).

8. The notions of *theoretical constraint* and *operational constraint* form part of logical positivist thinking about scientific theories: they refer to restrictions on the "acceptance" of theories, in the sense that for logical positivists a theory can only be accepted if it satisfies both constraints—to an extent to be determined. Theoretical constraints refer to the formal properties that a theory must possess: for example, having a certain number of axioms, being elegant, consistent, plausible, functionally

simple, conservative, etc. (cf. Putnam 1981: 31). Operational constraints prescribe the executability of certain operations and experiments if it follows from a given theory that they can be performed: operational constraints therefore connect the theory to the world, seeking to test its experimental adequacy (cf. Putnam 1981: 29ff.). Putnam extends the use of these constraints by applying them to possible "interpretations" of theories: since not every interpretation is as good as every other, these constraints can help narrow the class of permissible interpretations (see below).

9. M' is a submodel of a model M if its domain is a subset of that of M, and it assigns to predicate symbols the same relations that M assigns.

10. The entire second chapter of Putnam (1981) plus the *Appendix* are devoted to demonstrating the fact that "even if we have constraints of whatever nature which determine the truth-value of every sentence in a language *in every possible world*, still the reference of individual terms remains indeterminate" (Putnam 1981: 33).

11. One might ask what is the difference between ideal theory and OTT. For the metaphysical realist who believes in its existence, two possibilities open up: either the OTT is never formulated, even in principle, given the gap between the world and our cognitive faculties, or it may be the ideal theory itself. In the latter case, by virtue of the same gap we may never realize it, believing (erroneously) that it may be false.

12. God, Putnam maintains, could formulate the ideal theory. If He did, it would be a theory that humans, examining it, would accept, if they want to behave rationally. Humans, that is, would be able to appreciate its qualities and thus to accept it. And this is the sense of "ideal" at issue here (cf. Putnam 1980: 12).

13. This constraint concerns the objects that play a role in explaining our experience, and is understood by Putnam as a kind of *Ockham's razor*. Indeed it dictates that entities should not be multiplied beyond necessity (cf. Putnam 1981: 133).

14. The results in question are the theorem, proved by Gödel in 1930, that every consistent first-order theory has a model with a denumerable domain, and a corollary of it, that states that every consistent first-order theory has a model of cardinality α, for every cardinal number $\alpha \geq \aleph_0$ (where $\aleph_0$ indicates the cardinal number of every denumerable set) (cf. Mendelson 2015: 88 and 90).

15. The literature on Putnam's model-theoretic argument is enormous, more or less divided between commentators who are sympathetic and commentators who think that the argument is question-begging and does not succeed in refuting metaphysical realism. To get a rough idea of the debate, I suggest reading at least Lewis (1984), García-Carpintero (1996), Hale & Wright (1997b), van Fraassen (1997), Dümont (1999), Douven (1999), Hodesdon (2018), Bays (2001), Soysal (2020).

16. The arguments of Richard Boyd, Michael Devitt, Clark Glymour, Daniel Dennett, and David Lewis—while not all traceable within the framework of reductionist physicalism—are considered by Putnam in one way or another to be grist for the metaphysical realist's mill. Quine—"a physicalist but *not* a realist" (Putnam 1984a: 90): he "rejects metaphysical realism and the idea of a unique 'correspondence' between our terms and things in themselves" (Putnam 1983c: 223)—is left out of this group.

17. This Millian concept was taken up by John Mackie (1974: 60–64). As for post-Einsteinian physics, "we might define the total cause at time t_0 of an event A

(at a time subsequent to t_0) to be the entire three-dimensional space-time region that constitutes the bottom of A's light cone at the time t_0. Any aspect of this region that is sufficient to produce A (at the appropriate time t_1) by virtue of the Dirac equation (or the appropriate equation of motion) may also be called a 'total cause' of A" (Putnam 1984a: 87).

18. Cf. "what we call the 'cause' of what depends on more than the intrinsic physical properties of the objects and events concerned; speaking of causality is a way of introducing an *explanatory structure*, and [. . .] the same events get structured differently when different interests are brought to them (which is not to say that any event can be structured in just any way, of course)" (Putnam 1992n: 378).

19. "Science as we know it has been anti-metaphysical from the seventeen century on; and not just because of 'positivistic interpretations'. Newton was certainly no positivist; but he strongly rejected the idea that his theory of universal gravitation could or should be read as a description of a metaphysically ultimate fact. [. . .] And Newton was certainly right. [...] The physics that has replaced Newton's has the same property. [. . .] Worse still, from the metaphysician's point of view, the most successful and most accurate physical theory of all time, quantum mechanics, has *no* 'realist interpretation' that is acceptable to physicists. It is understood as a description of the world as *experienced by observers*; it does not even pretend to the kind of 'absoluteness' the metaphysician aims at" (Putnam 1983c: 227–228).

20. "Peirce also mentions the time-reversibility of the 'causal' relations of classical physics as a reason for doubting that what we have in the classical physical picture is 'causation' in the intuitive sense" (Putnam 1992n: 379).

21. An important clarification. In the chapter on Putnam's semantic conception, we saw that central to it is the thesis that once we have discovered what a certain natural kind (water, for example) is in the actual world, we have discovered its "nature": does this mean that for Putnam natural kinds have *intrinsic* properties? *Essences*? The answer is that we can in this case speak of "essentialism," but it is not the kind of essentialism cherished by the metaphysical realist, since Putnam stresses the role of *human referential intentions* in identifying what is to count as water: "the 'essence' of water in *this* sense is the product of our use of the word, the kinds of referential intentions we have: this sort of essence is not 'built into the world'" (Putnam 1983c: 221). Hence it follows that "the notion of 'causal chain' involved is that of an *explanatory* chain" (Putnam 1983c: 213). See above the last section of Chapter 4.

22. Both the essays in Goldberg (2016) and Thorpe & Wright (2022) are useful descriptions of the various perspectives from which the BIVs case advanced by Putnam has been discussed. Their reading is highly recommended.

23. It is interesting to notice the consonance between this idea and a note Wittgenstein wrote a couple of days before he passed away: "'But even if in such cases I can't be mistaken, isn't it possible that I am drugged?' If I am and if the drug has taken away my consciousness, then I am not now really *talking* and *thinking*. I cannot seriously suppose that I am at this moment dreaming. Someone who, dreaming, says 'I am dreaming,' even if he speaks audibly in doing so, is no more right than if he said in his dream 'it is raining,' while it was in fact raining. Even if his dream were actually connected with the noise of the rain" (Wittgenstein 1969: §676; my italics).

24. There is also a shorter version that is due to Crispin Wright at Putnam's own suggestion: "(1) In vat English, 'vat' does not refer to vats [first premise]; (2) In my language, 'vat' refers to vats [second premise]. Therefore, my language is not vat English—that is, I am not a BIV" (cf. Putnam 1992f: 404). However, the conclusion merges Putnam's (3) and (4); therefore, the outright shorter version is that of Tim Button: "A BIV's word 'brain' does not refer to brains. (2) My word 'brain' refers to brains. (3) So: I am not a BIV." (Button 2013: 108) Equally short is Thomas Tymoczko's version: "(1) We can raise the question: 'Are we BIVs?' (2) If we were BIVs, we could not raise that question. Therefore, (3) We are not BIVs." (Tymoczko 1989: 281) Gary Ebbs (2012: 28, 2016b: 34) also uses this version, albeit slightly modified. It is interesting to note that the differences from the previous ones are that here there is just one premise (the *causal* one), there is no (explicit) mention of languages, and the level at which the proof is placed is *epistemological*—the mentioned question is like those usually raised in the context of inquiry.

25. Obviously coming to know that we are BIV, if we are, is *genuine* knowledge. Cf. Putnam (1989c): 110–111.

Chapter Seven

Natural Realism

The period in which Putnam subscribed to internal realism can be dated precisely: in fact, it runs from 1976—the year in which he delivered the Boston lecture entitled "Realism and Reason" published in 1977 and that later became the last chapter of Putnam (1978a)—to 1990—the year in which he had the opportunity to present some new developments in his thought at a conference (the *Gifford Conference*) at the University of St. Andrews devoted entirely to his philosophy. In particular, it was in his response to Simon Blackburn's talk that the first public admission of change appeared (the written version would later appear in Putnam 1994f).

Constantly driven by a concern to provide a plausible explanation of the way language and thought "hook on" to the world, Putnam gradually begins to notice the flaws in his epistemic version of realism. To begin with, he decisively rejects the debut metaphor in Putnam (1981), the one that "the mind and the world jointly make up the mind and the world":

> I should not have seen us as "making up" the world (not even with the world's help); I should have seen us as *open* to the world, as interacting with the world in ways that permit aspects of it to reveal themselves to us. Of course, we need to invent concepts to do that. There is plenty of construction activity here, but we do not construct reality itself. [. . .] what we actually make up is not the world but language games, concepts, uses, and conceptual schemes (Putnam 2012c: 61–62 and 64; cf. also Putnam 1992f: 368).

The rejection of the metaphor with which Putnam characterized internal realism coincides with the rejection of *verificationist semantics*—the semantics on which internal realism rests. The justification of the correctness of our statements under sufficiently good epistemic conditions is after all achieved through our cognitive faculties and what they are able to verify: it is this kind of justification (empirical and otherwise) that, at the same time, fixes the extension of the truth predicate, establishes which statements are meaningful, and delineates the contours of the world (the world of the internal realist not being independent of *all* verification).

The adoption of this kind of semantics shows how strong an influence Michael Dummett had on Putnam around the mid-1970s.

In his reflections on the proper procedure on the basis of which to construct a theory of meaning, Dummett emphasized how an appropriate linguistic use by a speaker depends intimately on what s/he *knows* about language, depends on how s/he *understands* each fragment of the language s/he employs, where such understanding consists precisely in knowing the meanings of those fragments. Thus, the theory of meaning is identified with a theory of understanding (cf. Dummett 1975, 1976, 1978, 1991). His central notion, therefore, the one around which to assemble the various pieces of the theory, cannot be for Dummett a *non*-epistemic notion of truth, as so many philosophers—at least from Gottlob Frege to Donald Davidson—have held. For if the meaning of a statement is given by the conditions under which it is true, and if one interprets "true" on the basis of a non-epistemic notion, those conditions may or may not be fulfilled quite independently of what any speaker may know about it,[1] with the result that the latter will be deprived of any element that might lead her to understand that statement itself. Hence the Dummettian idea that the semantics for our language can only be of the verificationist type, the only one capable of ensuring that the whole process of understanding is under epistemic control.

It is this idea that Putnam takes on board and, in an essay written in 1976 ("Reference and Understanding," later published as the third chapter of Putnam 1978a), elaborates by developing the thesis that "the theory of language understanding and the theory of reference and truth have much less to do with one another than many philosophers have assumed" (Putnam 1978a: 97), a thesis designed to delineate a verificationist semantics for understanding and a truth-functional semantics for explaining the success of behavior (in general, and of scientists in particular). This is a thesis that later, in full internal realism, Putnam would reject by judging it impossible to hold two opposing semantics simultaneously, choosing the verificationist one and, we have been seen, proposing

> to identify being true not with being verified, as Michael Dummett does, but with being verified to a sufficient degree to warrant acceptance under sufficiently good epistemic conditions (Putnam 1994e: 17).

It is this identification that delineates the content of truth and thus of the world.

However, Putnam now realizes, the belief that an "access" to the world could only be delineated on the basis of our cognitive faculties and what they are capable of verifying, albeit under epistemic conditions judged to be good or ideal (to a varying degree depending on the statement under consideration), was but a hope destined to founder. Just as for the metaphysical realist there is the problem of explaining how our minds are able to connect to a world sharply separate from the minds themselves, so for the internal realist there is "an equal problem as to how we can have referential or other access to 'sufficiently good epistemic situations'" (Putnam 1994e: 18; and Putnam 2012c: 59). The identification of the reason for the impossibility of offering a plausible image of how the mind accesses the world—an impossibility common to both kinds of realism previously embraced by Putnam—along with a remedy intended to eradicate that reason at its root, constitutes the heart of Putnam's new realism: *natural realism*.

THE SOLIPSISTIC TRAP

We can now take up the "but" left above. Putnam had long thought that the position of the metaphysical realist did not make sense; *but* he had not considered

> the possibility that it was internal realism that made no sense, or, more precisely, that it made sense only if one was prepared to retreat all the way to solipsism (Putnam 2012b: 78).

What solipsism has to do with it is quickly said. If my understanding of the statements of my language is given by *my* ability to verify them, then any statement about the world, and about others as part of the world, can only be understood by *me*: my understanding of a statement is reduced to my ability to say what confirms it, and to what degree, with the consequence that I can only access ideal epistemic conditions solipsistically. It would be useless to retort that verificationism, if we are to understand it correctly, can only be "intersubjective," in the sense that verifications presented for or against a statement can only be public in order to have value and thus be accepted, constituting the meaning of the statement in question: the fact is that "if my understanding of the counterfactual '*S* would be justified if conditions were good enough' is *exhausted* by my capacity to tell to what degree it is justified to assert it [then not only] my understanding of *S* is just my capacity to tell what confirms *S*" (Putnam 2012b: 79), but my own discourse about others and their verifications "is only intelligible to *me*" (Putnam 2012b: 79).

Internal realism thus loses all aspiration to some intersubjective notion of truth and,

> far from being an intelligible alternative to a supposedly unintelligible *meta-physical realism*, can itself possess no *public* intelligibility [. . .] If this is right, then it clearly becomes vital to find an account of our capacity to understand and use language that "fits" with realism (Putnam 2012b: 80).

Not only (radical) metaphysical realism but also internal realism thus proves meaningless: it is not the ideal epistemic conditions that can provide access to the world by virtue of a genuine understanding of the statements of our language.[2] In other words, this is not how the mind is able to connect to the world. But why?

THE INTERFACES BETWEEN US AND THE WORLD

In addition to the unavoidable Dewey and Wittgenstein, William James, and J. L. Austin—and to some extent John McDowell—were also exerting a decisive influence on Putnam's thinking: authors whose writings Putnam had been visiting for decades but who were only now being fully appreciated. Of the various aspects of Austin's thinking, Putnam focuses on the critique of sense data which—according to a long tradition of thought—were supposed to be the primary object of our perceptual experiences of the external world (cf. Austin 1962). This critique makes it clear to him that "the 'how does language hook on to the world' issue is, at bottom, a replay of the old 'how does perception hook on to the world' issue" (Putnam 1994e: 12).

The traditional idea that the perception of an object takes place because of a myriad of sense data that gives us information about the object's multiple features, and that therefore what we are directly connected with is not the object itself but is sense data, configures the same epistemological difficulty in which—after all—Descartes was trapped: that of the opposition between a mental substance and a physical substance so sharply separated as to admit the hypothesis that we might be BIVs. What could possibly ensure that the cause of the sense data I perceive is some object existing in the world and not the computer software connected to BIVs? And even in the case that we are not BIVs, what could ensure that the sensory data give us a faithful representation of the object and are not instead distorting the perception itself in inscrutable ways?

If we remain, therefore, within the traditional coordinates of perception, the appeal to pernicious interfaces between us and the world, the face of which is that of sense data, seems inevitable. And not only sense data, but also conceptual schemes:

> on the "internal realist" picture it is not only our experiences (conceived of as "sense data") that are an interface *between* us and the world; our "conceptual schemes" are likewise conceived of as an interface. And the two "interfaces" are related: I saw our ways of conceptualizing, our language games, as controlled by "operational constraints" that ultimately reduce to our sense data (Putnam 2012c: 61).

Hence the solution appears simple and consequential: to adopt an account of perception that excludes any possible intermediary between us and external reality, along the lines of James, Austin, and McDowell,[3] an account that can form the core of a metaphysical position more faithful than the previous two to the actual human condition and which, borrowing a formulation "from William James's expressed desire for a view of perception that does justice to the 'natural realism of the common man'" (Putnam 1994e: 10), is called *natural realism.*[4]

MIND, PERCEPTION, UNDERSTANDING: LIBERAL FUNCTIONALISM

The perceptual relationship that the human mind is able to entertain with objects in the surrounding environment is now regarded by Putnam as *direct*, that is, devoid of intermediaries intended to mediate between a mind and a world regarded as separate.[5] In the course of perception, between the human mind and its surroundings there is no separation to be bridged by a third element, and consequently the problem of how these two manage to enter into relationship does not even arise: if there is perception, then the relationship is guaranteed. But—Putnam is keen to point out—the problem does not arise so long as one presents a conception of mind that departs from the conception inherited from Descartes and in various ways revitalized by the Western tradition: a revision of the way the mind perceptually connects to the world cannot therefore fail to *bring with it* a revision of the conception of mind itself.

In this way, a mind understood as a thing, an object, an organ, an entity in itself enclosed and self-sufficient because it already possesses all its powers, a mind identifiable (in the absence of better alternatives) with the brain, gives way to a mind understood as a system of theoretical and practical skills that is refined over time but involves the world and its objects from the beginning:

Mind talk is not talk about an immaterial part of us but rather a way of describing the exercise of certain abilities we possess, abilities that supervene upon the
activities of our brains and upon our various transactions with the environment
but that do not have to be reductively explained using the vocabulary of physics and biology, or even the vocabulary of computer science (Putnam 1994e:
37–38).

The combination of this conception of mind and this conception of perception constitutes the account of us human beings most in keeping with
Putnam's new realism. Indeed, one would not do justice to the direct nature
of the perceptual link we have with the world if one were to place this direct
and immediate link within the traditional conception of the mind as an entity
already complete and in possession of all its powers. If one continues to
accept this traditional conception, and explains on that basis how some object
can enter into a perceptual relation with the mind (an obviously causal relation), it becomes inevitable to regard the causal link as embedded in the structure of reality. Resting on an implicit dichotomy between subject and object,
internal and external, the traditional conception after all gives us an image of
mind and world as *separate* entities, where any nexus that unites them can
only be external just as everything that is not mental is external. Therefore,
the causes of our perceptions may well be considered "direct," but they will
still be "external," just like the world as a whole: exactly the "non-Humean"
conception of causation identified by Putnam as a misplaced attempt to substantiate metaphysical realism. Such causes—substantiating a non-cognitive
connection to objects—motivate the introduction into the philosophical
lexicon of expressions such as "'external' things [*cause*] 'experiences,'" conceived as affectations of our subjectivity, which is what qualia are conceived
to be" (Putnam 1994e: 20), expressions that sanction the belief in the existence of a hiatus between mind and world by making it impossible to identify
a way to connect them. But, Putnam now points out, "sensory experiences
are not passive modifications of an object called 'mind'" (Putnam 1994e: 37).

The mind conceived as a system of concrete abilities in direct contact with
the world thus appears to be an indispensable element of natural realism. Not
only that: such a conception of mind allows Putnam to amalgamate some of
the pieces of his philosophical position more satisfactorily than before.

In Chapter 3 we saw why Putnam—while still subscribing to the idea that
the human mind should be identified on the basis of the functions it per-
forms—repudiated his initial version of functionalism: the reason certainly
lay in the realization that the mind is not only "compositionally" plastic, but
also "computationally" plastic; just as certainly, however, that reason lay in
the rejection of an assumption implicit in that kind of functionalism, namely,
that the human mind has within it all the resources to generate thoughts and

statements already endowed with a reference to the external world, even if these resources are described solely in terms of aspects of the internal "software," without any specification of how the internal-external relationship should be secured. It is true that in the early 1970s, a few years after popularizing his own functionalist hypothesis, Putnam was beginning to develop as a philosopher of language a reflection around meaning that had a clear and distinct externalist character, but the fact remains that he failed adequately to decant this move into his philosophy of mind. Functionalism therefore seemed destined to remain a brilliant idea but substantially at odds with the fundamental lines of his thought.

And now, in the early 1990s, Putnam identified the basic flaw in his initial functionalism: the assumption of the existence of "sense data," the supposed intermediaries of perception. Several traces of this presupposition can be found in the writings of that period; here is one:

> although the philosopher John Austin and the psychologist Fred Skinner both tried to drive sense data out of existence, most philosophers and psychologists think that there are such things as *sensations* or *qualia*. They [. . .] may be somewhat ill-defined entities rather than the perfectly sharp particulars they were once taken to be; but it seems reasonable to hold that they are part of the legitimate subject matter of cognitive psychology and philosophy, and not mere pseudo-entities invented by bad psychology and bad philosophy (Putnam 2012c: 60, which picks up verbatim from Putnam 1980: 15).

Putnam is now well aware of just how bad a certain kind of psychology and a certain kind of philosophy—those that presuppose sense data—can be. The renewed conception of mind thus allows him to recover functionalism by emphasizing its anti-reductionist elements: this is what makes him—as he would declare years later—a *liberal functionalist*:

> What I have in mind in speaking of a "liberal functionalist" is someone who, like me (or like me today), accepts the basic functionalist idea that what matters for consciousness and for mental properties generally is the right sort of *functional capacities* and not the particular matter that subserves those capacities, but (1) does not insist that those functions be "internal," that is, completely describable without going outside the organism's "brain" (thus Gibsonian "affordances" and Millikan's "normal biological functioning" in an environment can all be involved in the description of the "functional organization" of an organism); (2) does not insist that those capacities be described as capacities to *compute* (although she is naturally happy when computer science sheds light on some part of our functioning); and (3) does not even eschew intentional idioms, if they are needed, in describing our functioning (Putnam 2012b: 83).

The elimination of sense data from the description of the perception and functioning of the human mind thus allows for the elimination of "the model of the mind as something 'inside' us" (Putnam 1992g: 357) and of language "inside" the mind, resulting not only in a revision of functionalism, but also of the way in which we understand *language understanding*.

From the earliest reflections on language, and then gradually in the writings of the 1970s collected in Putnam (1978a), in its broad outlines such understanding was explained on the basis of the mastery of the "use" of language. However, the strong conditioning placed by early functionalism on the interpretation of any mental function had led Putnam to regard speakers' use of linguistic expressions as "a 'cognitive scientific' notion, that is, use was to be described largely in terms of computer programs in the brain" (Putnam 1994e: 14). It is true that there was more than just this, since his semantic conception showed the indispensability of a consideration of the environment in which minds and their language are situated, the environment whose objects and events "cause" speakers to use certain expressions; but, we have seen, within the framework of the inside/outside distinction typical of the tradition inherited by Putnam, whatever lies beyond the sense data perceived by the brain "is connected to our mental processes only causally, not cognitively, [since] everything outside our skin is also outside our cognitive processing" (Putnam 1994e: 16). Causes, objects, and events external to all human cognition can only be considered part of a hypothetical structure of a world "in itself," exactly the kind of world denied by Putnam as meaningless. The central role of the world was therefore not being given justice—philosophical justice—it was a role that was recognized only in a Pickwickian sense, as ultimately shown by the model-theoretic argument, the one that leads to absolute Skolemization: different interpretations of our language can agree on what sense data are associated with certain linguistic expressions, while diverging radically about what they refer to. And appealing to usage by remaining within this framework solves nothing: to assert—as Putnam of the late 1970s put it—"either the use *already* fixes the 'interpretation' or *nothing* can" (Putnam 1980: 24), by virtue of the fact that we conceive (implicitly and unconsciously) of use as determined by causes belonging to a world outside the sphere of cognition, outlines an unlikely "world that interprets our words for us, a world in which there are, as it were, 'noetic rays' stretching from the outside into our heads [. . .] a magical world, a world of fantasy" (Putnam 1994e: 17). A world so far removed from what the human mind can comprehend, that it evaporates into absolute insubstantiality.

Now, it is undeniable that nothing but use can fix the interpretation of our words, but this Wittgensteinian perspective gains depth only if the use is that of a mind with "long arms" (Putnam 2012b: 83), that is, in direct cognitive contact with the world: precisely the kind of mind that, as we have seen

Putnam now argues, is not an organ but a "system of abilities." Understanding a language thus means *having the abilities that one exercises when and in using language* (Putnam 1994e: 15), just as

> learning to use words is more like learning to play a musical instrument than like learning to extract square roots. Sensitivity is involved, and so is informal rationality, and there is room for individual creativity (Putnam 1994f: 243).

Learning and understanding a language are such varied tasks that they elude any attempt to harness them in the net of a narrow set of rigid, predetermined rules, just as it is impossible to harness human creativity on the basis of one or more algorithms. These are complex activities that involve a person in his or her entirety: intellectual abilities, practical skills, intuition, and common sense. When we are able to understand a word or statement, what has happened is not a mere association of meanings to linguistic elements that present themselves initially as mere marks on paper or sounds passing through the air, an association enabled by some automated procedure in the brain—innate or learned. Words and statements already have semantic value because they have already entered into the interactive and cognitive circuit between mind and world, a circuit in which mind and world are already in immediate connection. So much so that

> when we know and use a language well, when it becomes the vehicle of our own thinking, and not something we have to mentally translate into some more familiar language, we do not, *pace* Richard Rorty, experience its words and sentences as "marks and noises" into which a significance has to be read. When we hear a sentence in a language we understand, we do not associate a sense with a sign design; we perceive the sense *in* the sign design. Sentences that I think, and even sentences that I hear or read, simply do refer to whatever they are about—not because the "marks and noises" that I see and hear (or hear "in my head," in the case of my own thoughts) intrinsically have the meanings they have but because the sentence in use is not just a bunch of "marks and noises" (Putnam 1994e: 46).

Putnam goes on to reiterate here that the linguistic expressions we use and understand do not have *intrinsic* meaning, as if a semantic valence were attributed to them by some mysterious rigid and prefixed structure of the world; rather, they acquire meaning and reference thanks to us, to the abilities we display when we use language on a collective level, foremost among them our ability to perceive in a *broad sense* and in a *narrow sense* the meanings the world offers to our linguistic expressions. In the "broad sense" because "we perceive the sense *in* the sign design," and in the "narrow sense" because even the understanding of common words such as "chair" or "cat" would not be possible if we did not have even a minimal perceptual ability to recognize an object as such. And the same is true for the statements we assert that pertain to the objects, animals and people we interact with on a daily basis. Indeed,

> the existence of statements of this kind is a conceptual prerequisite of our being able to understand a language at all (Putnam 1995: 299).

In other words, our whole linguistic competence rests on our perceptual interactions with the world, the interactions that allow us to justify the truth of statements about the empirical world.[6]

But—one will ask oneself not without some wonderment—is this not exactly what Putnam claimed when he was an internal realist? Perceptual justification is precisely what allows empirical statements to be *verified*, making their assertion guaranteed and thus legitimate. And if the general scaffolding of our language rests on this kind of justification, should we not conclude that, after all, truth is a kind of warranted assertibility under sufficiently good epistemic conditions? Does not the verificationism that, since the 1990s Putnam so carefully dismissed, thus come back in through the back door? It would seem that, after all, Dummett was right when he argued that the question of realism comes down to just two philosophical options:

> either our sentences have only assertibility conditions or they have something mysterious floating above them and connecting them with reality (Putnam 1994e: 46).

Dummett, as is well known, chose the former disjunct. His aversion to the traditional notion of truth—a "non-epistemic" notion—was motivated by the impossibility of providing an explanation of the understanding of statements based on truth conditions, since these, for certain statements, can radically transcend our cognitive capacity. Hence his anti-realism, since our notion of the world flows from the statements that can be considered true, and these for Dummett are identified with those that *we* can verify. However, according to Putnam, Dummett was not right: far from subscribing to the disjunction envisaged above, he breaks it down. Let us see why.[7]

THE UNDERSTANDING OF LANGUAGE

It is precisely the considerations around *use* that enable Putnam to break down the Dummettian disjunction and dispel the objections that so concerned Putnam "at the very beginning of what has proved to be a long journey from realism back to realism (but not [. . .] back to the metaphysical version of realism with which I started)" (Putnam 1994e: 49).

It should be noted that the considerations of use, enabled as we have seen by a renewed conception of mind, highlight one of the leitmotifs of Putnam's reflection: his respect for the verbal and non-verbal behaviors in which we are constantly engaged, and at different levels of complexity, a respect for "our most basic intellectual practices" that induces us to examine things "from inside what we normally regard as our best and most rational practice, without philosophical revisionism" (Putnam 1992f: 365). A respect that amounts to assigning central philosophical importance to *common sense*.

It is this respect for common sense that makes him say things like "I still continue, of course, to insist that our grasp of the notion of truth depends on our grasp of the notion of warranted assertibility" (Putnam 1992f: 365), because this is precisely what happens in so many instances of our ordinary, scientific experience. Following some of Gary Ebbs's remarks, Putnam subscribes to the idea that "there is an *interdependence* between the notions of truth and rational acceptability, because they are both rooted in the norms underlying our everyday and scientific inquiries" (Ebbs 1992: 26). Gone, however, is the typical belief of the internal realist that *everything* there is to be said about truth lies in the relationship between these two notions, since "we are virtually *never* in a position to equate the truth of a particular statement with its verifiability in situations which we can precisely describe" (Ebbs 1992: 26). Putnam is now persuaded that "there is no simple relation between the truth of a statement S made by a person P and the situations in which P would be justified in accepting S. [. . .] The most we can say is that our *understanding* of what it means to say that a statement S is true is essentially tied to our conception of situations in which S would be correctly affirmed by someone or other" (Ebbs 1992: 25–26; italics mine).

Yet, Putnam does not fully identify with these theses of Ebbs: although he believes them to be correct and indispensable for obtaining a faithful notion of what truth is, he says that Ebbs's last sentence in the quotation just quoted "may be too strong" (Putnam 1992g: 357). Although he believes that there is a close connection between truth and rational acceptability, he does not think, as he previously had, that the issue ends there: first, because our ordinary practice shows that we possess an ability to understand statements that concern the *past*, and thus situations for which we do not, or no longer, have empirical evidence on which to rely to justify these statements, that is, to verify them; second, because that same practice shows that we also possess an ability to understand statements that are unverifiable even in principle, and for which there is not and can never be any justificatory evidence.

As to the first point, if the verificationist were right, statements concerning events that happened in the past for which we possess no reliable recollection

or testimony amount to little more than a *façon de parler*: if there is no
evidential level on which to rest the assertion of a statement of this kind,
because this level has been lost, then—as the logical positivists argued—the
statement cannot be true or false and is devoid of cognitive value. Dummett,
for his part,[8] is known to have asserted that the only possible "account of that
in which our grasp of the truth-conditions of such sentences consists" (Dummett 1976: 62) works only if we are "capable of observing the past as we
observe the present." It is our implicit presupposition that according to which
"it is only a practical difficulty which impedes our determining the truth-
values of sentences" about the past, a presupposition that, "when challenged,
is defended by appeal to a hypothetical being [. . .] for whom the sentences
in question would *not* be undecidable" (Dummett 1976: 61), a superhuman
observer immersed in a spatiotemporal perspective different from our own,
"whose observational and intellectual powers transcend our own, such powers being modelled on those which we possess, but extended by analogy"
(Dummett 1976: 60). It is because he finds this position indefensible that—as
the disjunction above reveals—Dummett then develops an explanation of
understanding based on assertibility conditions.

Dummett's conclusion was guided by the belief in the existence of a regulative principle that we have already encountered in Chapter 6 and which he
believes guides the notion of truth: that "if a statement is true, it must be in
principle possible to know that it is true" (Dummett 1976: 61). Admittedly,
a *regulative* and not a *constitutive* principle: for Dummett it regulates our
verbal behavior and guides our intuitions about truth but it does not enter into
the "constitution" of truth, it does not help describe its nature. Be that as it
may, according to Putnam to hold that such a principle underlies our practice
is a clear sign of a lack of attention to that practice itself. The verificationist
approach collides with a datum easily found in our linguistic habit, a datum
that reflects "the commonsense idea that we can conceive of how [. . .] some
past event was, and not just in the sense of conceiving of what it would be to
verify *now* that that is how it was" (Putnam 1994e: 47). When, for example,
we assert a statement such as "More than one sniper shot John F. Kennedy
on November 22, 1963," we all grant that this statement has a definite truth-
value: either it is true or it is false. But we grant it while being well aware
that there is no overwhelming empirical evidence in favor of truth or falsity;
while being well aware, therefore, that any possibility of verification is ruled
out both now and in the future. Therefore, in understanding that this statement
is true or false in a genuine, full sense, we show that we *understand* the statement independently of any possibility of verification: we understand it on the
basis of its *truth conditions*, not on the basis of its verification conditions. In
fact, we certainly cannot mean that it is possible to verify it *now*, since we are
talking about a fact about which there is no evidence at present, but neither

are we implicitly saying that, if we had a sufficiently long time to carry out investigations, we would succeed in verifying or falsifying the statement, that is, we are not alluding to some ideal condition obtainable in the *future*—perhaps because we will be able to gain the optimal epistemic situation of a superhuman being in the future. With firm resolve, Putnam declares that "any theory that makes the truth or falsity of a historical claim depend on whether the claim can be decided in the future is radically misguided" (Putnam 1992g: 357), just as Peirce was in error when he claimed that by pursuing scientific research steadfastly and stubbornly, *in the long run* all problems will be solved and we will be able to assign a truth-value to any statement: such a position is misleading because

> it assumes (1) that future time is infinite—something we are no longer willing to postulate; and (2) that no information is ever irretrievably destroyed—something that also contradicts our present physical theories (though not, of course, the physical theories of Peirce's time): Hawking has shown that there is irrecoverable loss of information in Black Holes (Putnam 1995: 295).

In other words, it belongs to our pre-theoretical intuitions that a statement concerning past events such as the one about John F. Kennedy refers to a fact of the matter, even if it is a fact that is *no longer* empirically accessible and therefore unknowable; such a statement is not a mere figure of speech, it is not a mere *façon de parler* empty of content since it is supposed to be the result of purely formal linguistic conventions, nor is it something verbally inert that can acquire a communicative force of its own because an interpretation is associated with it by means of "assertibility conditions." On the contrary, the very datum concerning our linguistic custom referred to above—a datum allowed by our language and which is therefore an exclusive heritage of our species—shows that this custom *already* benefits from all the substance that the world can give it, and that therefore language and the world are closely intertwined: it is true that "our highly developed and highly discriminating abilities to think about situations that we are not observing are developments of powers that we share with other animals, [but] our power of imagining, remembering, expecting what is not the case here and now is a part of our nature,"[9] and not that of other species, due to the fact that "language alters the range of *experiences* we can have" (Putnam 1994e: 48) and is "a way of extending our natural powers of observation" (Putnam 1994e: 56):

> the ways in which language extends the mental abilities that we share with other animals are almost endless; our ability to construct sophisticated scientific theories is only one example. A very different sort of example is provided by the role of logical constants, for example, the words *all* and *no*. An animal or a child that has not yet learned to use these words may have expectations that

we who have acquired them can and do describe with the aid of these words (Putnam 1994e: 57).

The idea that the use of logical constants—such as *all, no, not,* and so on—cooperate to enlarge our mental capacities brings us to the second point.

A careful consideration of certain aspects of our linguistic practice can help us highlight the fact that we are able to immediately understand even statements whose truth-value we could not verify even in principle. The difference with the previous case is that, whereas before our inability to know how facts had unfolded stemmed from the *contingencies of epistemic opportunity*, that is, from cognitive limitations that are not at all necessary but wholly incidental because they pertain to the specific and random "spatiotemporal situation, or perceptual or intellectual capacity" (Wright 2000: 360) with which we happen to be endowed—so much so that if we or others had been better positioned or more cognitively gifted we would have enjoyed that opportunity—in the present case, such epistemic opportunity is necessarily ruled out: not even a superhuman being endowed with the same observational and intellectual capacities as we are but extended as much as we want could achieve here any cognitive result.

There are several examples with which Putnam substantiates this point. One goes as far back as the 1960s when, as a naive metaphysical realist, he subjected the doctrine of logical positivism to a slashing critique. It is the example[10] involving

> the sentence "There is a gold mountain one mile high and no one knows that there is a gold mountain one mile high," [which] is, if true, unverifiable. No conceivable experience can show that both conjuncts [in the sentence] are simultaneously true; for any experience that verified the first conjunct would falsify the second, and thus the whole sentence. Yet no one has ever offered the slightest reason for one to think that [this sentence] could not be true in some possible world (Putnam 1969a: 443).

Although this is a purpose-built and, moreover, somewhat fanciful example, it provides evidence that there are situations that we are capable of conceiving and sentences that we are capable of understanding but which, nevertheless, we are unable to know and verify even in principle.

Another example used by Putnam, again purpose-built and imaginative, involves a particular polygon. It involves the statement "There is no physical system consisting of 100 stars arranged at the vertices of a regular 100-gon." As in the case of the gold mountain, the existence of this particular complex object is not forbidden by the physical laws currently known, and there is therefore a small probability that it can actually exist. In other words, the statement does not configure a physical impossibility, let alone a logical impossibility. However, although possible, it could be the case that that particular arrangement of stars does not exist anywhere in the universe at all—especially if the universe is finite. Therefore,

> it follows *from our current scientific world-picture itself* that there is no way we could know that this is the case if it is. After all, we cannot have any causal interaction of any kind with space-time regions outside our light cone (i.e., with regions such that a signal from those regions would have to travel faster than light to reach us) (Putnam 1995: 294).

So, here again we have a statement of which, even if it is true, it is necessarily impossible to know that it is so.

The fact that Putnam resorts to his imaginative vein to illustrate these kinds of statements may give the impression that, after all, this has little to do with our linguistic-cognitive situation. That would be a mistaken impression but, in any case, there is an example—which Putnam ends up favoring—that concerns a real cognitive situation. For decades, in fact, there have been space programs of the most diverse purposes, including the purpose of ascertaining whether there is any form of intelligent life in the universe other than that present on Earth. Well, a statement such as "There are no intelligent extraterrestrials" (let's call it *K*) is certainly part of the linguistic-cognitive practice prevalent among scientists and laymen alike, and—again—this statement, if true, is unverifiable: the situation it envisages, in the event that it actually does arise, is not knowable even in principle. In spite of the fact that

> there might be *overwhelming evidence* that there are intelligent extraterrestrials (somewhere, some time), *evidence* for laws according to which the probability that such never did, do not, and never will exist, is less than one in a trillion, let us say (which would certainly justify *believing* that intelligent extraterrestrials exist in spacetime), [it might instead be that,] in fact, ours is a universe in which that one in a trillion chance that they do not exist is realized (Putnam 2012b: 75).

Suppose then that this one in a trillion possibility has been realized, that it is a fact, and therefore that there are no intelligent extraterrestrials. This fact is unknowable, and it is so

> partly for logical reasons—the impossibility, for example, of verifying a negative existential statement like [*K*], in case it is true—and partly for empirical reasons (the inaccessibility of information from beyond the "event horizon," or from the interiors of black holes) (Putnam, 2012f: 100–101).

The empirical reasons are exactly those seen regarding the statement about stars arranged as if each were at the vertex of a regular polygon of 100 angles, those reasons why,

> since information cannot travel faster than light, most parts of the universe are sufficiently far away that causal signals from them showing that they contain no such extraterrestrials could never reach us. Thus it may be physically impossible for human beings ever to know the truth that intelligent extraterrestrial do not exist, if they don't (Putnam 2008c: 89).

Logical reasons, on the other hand, have to do with the obvious impossibility of verifying any universal statement such as "All crows are black" (since this statement concerns the totality of crows, and since the set of crows is *indefinite*, it cannot be ruled out that at a certain time or in a certain place it may turn out to be possible to observe a non-black crow). The same applies to the negation of an existential statement such as K, which will become clear if one considers that its form, $\sim\exists x\, Px$, is logically equivalent to $\forall x \sim Px$, that is, to a universal statement.[11]

Here, then, is an example of a statement that, if true, is unknowable as such—forever and ever.

Now, keeping in mind the Dummettian disjunction noted above, let us ask how it is that we "understand" a statement such as K. A verificationist *à la* Dummett would say that "it is just a *grammatical illusion*" (Putnam 1992f: 365) that K is unverifiable, if true. Since we understand it, s/he would argue, this means that there are conditions for its assertibility, and these conditions could be satisfied by the existence of some *physical reason* that prevents the formation of intelligent life forms in places other than Earth—say, a hitherto undiscovered law of nature. We can then say that K is true insofar as the statement "There *cannot* be intelligent extraterrestrials" is true, let us call it J, where the truth of J would be precisely guaranteed by that physical reason. This, the verificationist would continue, is the only way in which we can say that K is true, a way that emphasizes that K is verifiable via the verification of J. Indeed, if that physical reason exists, then it is after all fair to think that it is possible to discover it in principle. And s/he would conclude by reiterating that this is the only way to explain how we understand K because, otherwise, one would be forced to appeal to mysterious truth conditions.

However, Putnam retorts, it is counterintuitive that underlying our understanding of K is a line of reasoning, albeit implicit, such as that assumed by the verificationist. Things are actually much simpler, especially if we consider that K can be true in the complete absence of a physical reason that makes it so: as in the case of the statement about the constellation representing a regular 100-gon,

The idea underlying metaphysical realism, the idea that the world is already ready-made, that it contains a predetermined set of objects and an equally predetermined set of properties or relations that these objects may possess or find themselves in, the idea that the world is totally detached from and independent of the human mind and language, which, in their cognitive endeavors, attempt to reach it, facilitates an image that what is human and what is not human are by constitution rigidly separate, and come into contact—assuming they succeed in doing so—by virtue of a metaphysical relationship of correspondence "external" to both language and the world. It is the often-uncritical image of traditional realists that can only make sense when contemplated from a strictly external, objective, absolute, non-human point of view: the point of view of the God's Eye from which to observe how human and non-human come into contact. Since immobility and predeterminacy are part and parcel of such an image, from the philosophical impracticality of the latter follows the belief that reality is something plastic, malleable, constantly evolving, endowed with "many levels of form, including the level of morally significant human action" (Putnam 2008a: 5–6), a plural reality in which no one level has an ontological privilege that allows it to form the basis for any reduction of the other levels. Each level is singled out because of the role it plays in our lives, and it retains ontological dignity as long as it plays that role. It follows that, in addition to the moral level just mentioned, the material, aesthetic, religious, economic, legal, mathematical levels are part of the world, and the same applies to all those aspects of reality that arouse in us a perceptual or intellectual response going to constitute what—in the broadest sense of the term—is our *experience*.[13]

With respect to each of these levels one can speak of "correspondence to the facts" in the light and non-metaphysical sense just indicated: each of them presents such characteristics as to guarantee the objective validity of the statements concerning them, and if any statement is objectively valid, this is so because there is some aspect of reality to which that statement refers and which is responsible for that validity. It is therefore fair to claim that a statement is true insofar as it "corresponds" to some aspect of reality (the quotation marks here are to indicate the metaphysically light sense of correspondence clarified above). And it should be reiterated that this is true at *every* level, precisely because we are dealing with levels of *reality*. Reality impacts on us in a variety of ways, perceptual and otherwise; it *demands* something of us, challenges us by presenting us with problematic situations and prompting us to respond to stresses that occur not only on a physical level, but that invest our human nature in all its complexity. This is why, according to Putnam, reality is not morally, legally, aesthetically or religiously inert. This becomes clear when we reflect on the fact that the values we fix to guide our verbal and nonverbal behavior are of course created by us, but in response to needs that we do not create and that then guide us to evaluate which values are right and which are not, which of our responses are correct and which are not, bringing about a more or less profound process of revision of our value system, the beliefs we base on it, and the behaviors guided by those beliefs.[14] And it is precisely because this process, at all levels of reality, is virtually endless (and because world and language are inextricably intertwined) that reality is plastic and constantly changing.

Among the reasons that make the process of revising our general system of knowledge endless is that we cannot endow our knowledge with a solid and unbreakable foundation: as human beings, such a foundation is not among the things to which we can aspire, and this—though not only this—makes for an endless inquiry. There is nothing wrong with that: on closer inspection, such a foundation for our intellectual and practical existences is not at all indispensable: "as Peirce once put it, [. . .] we are on swampy ground, but that is what keeps us moving" (Putnam 2002a: 102). At the same time, given the intertwining of language and the world, given the intertwining of fact and convention, and given the fact that the world takes on epistemic visibility through true statements, it is the same continuing motion of inquiry that testifies to the magmatic and kinetic character of reality—and vice versa. Indeed,

> an essential part of the "language games" that we play in science, in morals and in the law is the invention of new concepts, and their introduction into generalized use; new concepts carry in their wake the possibility of formulating new truths (Putnam 2002a: 109).

And such new truths unveil new portions of the world, and new portions of the world enable the formulation of new truths: this is why "we endlessly renegotiate—and are *forced* to renegotiate—our notion of reality as our language and our life develop" (Putnam 1994e: 9).

This very ceaseless renegotiation of the notion of reality shows how unwarranted is the belief—implicit in the metaphysical realism from which Putnam had started—that the notions of "object" and "existence" are fixed and given once and for all. On the contrary—and this is the heart of the conceptual relativity we saw in Chapter 2—the variegated uses of language that we deploy in the most diverse contexts of everyday life stand to show that these notions possess a characteristic *openness*, by virtue of which what counts as an object and what counts as "existing" is not univocally predetermined from the outset. Of course, we generally use the words "object" and "exists" to refer to the medium-sized physical objects that populate the environment in which we live, and we can regard this as the initial use of these words, but the fact that we then talk about such things as World War II, the color of the sky, an image reflected in a mirror, the objects of desire, the number two, the ideal of freedom, the value of consistency, the main character of the last novel we read, and so on, is a clear sign that "object" and "exists" are being adapted to a plurivocity of uses that have almost nothing to do with the initial use.

The ontological commitment we make through our actual linguistic exchanges is thus not univocal: we manifest commitments of different kinds precisely because we make different uses of words such as "object" and "exists,"[15] and it is this that the wide range of our ontology unfolds. Not only that, it is an ontology that expands as our notion of object expands, and both expand in direct proportion to the growth of knowledge. A striking example is offered by physics:

> the reason that the quantum mechanical "particles" are not objects in the traditional sense is that in contemporary quantum mechanics particles have no definite number at all (in most "states"). [. . .] And the logical properties of quantum mechanical "fields" are equally bizarre. [. . .] Thus quantum mechanics is a wonderful example of how with the development of knowledge our idea of what counts as even a *possible* knowledge claim, our idea of what counts as even a *possible* object, and our idea of what counts as even a *possible* property are all subject to change (Putnam 1994e: 8).

The first three assumptions of metaphysical realism listed at the beginning of Chapter 6 thus seem completely unwarranted: words like "object," "entity," "being," "exist," and the like are not unchanging monoliths sheltered from the relentless growth of human knowledge, but on the contrary possess "an ever-expanding open family of uses" (Putnam 1994f: 243), an expansion amplified by that conceptual relativity that is a cornerstone of Putnam's thought. The fourth assumption also collapses as unjustified, as we have abundantly ascertained: there is no well-determined correspondence relation capable of explaining the truth of all statements of a language, regardless of their content, so as to formulate a general "definition" of truth embedded in a specific theory. And this is not so much because it is possible to give at most "a picture of our notion of truth (I doubt we can give anything that deserves the title of a 'theory')" (Putnam 2013a: 98), but because even in the case most favorable to the correspondentist conception—that of empirical statements—there would be a plurality of types of correspondence, and not just one as that conception would like:

> "This piece of beef weighs one pound" may "correspond to a reality" by the standard of correspondence appropriate to a butcher's shop, but be extremely wrong by the standard of laboratory science. And the kind and degree of correspondence changes again when the statement is "John is very neurotic" (Putnam 2013a: 98).

Moreover, believing that one can identify truth with a metaphysically strong correspondence relation might work in the case of some empirical statements, but it would be completely counterintuitive for understanding how statements that pertain to levels of reality other than the empirical level manage to be true: for example, it is impossible on this basis to explain the truth of statements concerning the ethical, mathematical, logical, legal, aesthetic, etc., levels. Consequently, the collapse of the fourth assumption requires us to provide an interpretation of the concept of truth that adheres to our actual metaphysical and cognitive situation: what, then, is the picture of truth coupled with natural realism?

ALETHIC PLURALISM

Here the full influence of the later Wittgenstein on Putnam's thought stands out.[16] Not only because of the conviction that our notion of reality, reality itself, and our language are constantly evolving, and not only because of the conviction that employing a language is immediately intertwined with reality—an intertwining that shows once again the "direct" relationship between mind and world—but also because of the emphasis on the breadth and variegation of uses, contexts, criteria, and parameters involved in speaking: "Wittgenstein was right in saying that language is a motley, *in the sense that* we have many different standards for different types of discussion" (Putnam 1978a: 116), for different types of statements and different types of words—including the word "true."

As for statements, we have seen that Putnam's account of understanding is based on the idea that they possess genuine truth conditions, and that to identify what these are it is sufficient to exemplify the *equivalence thesis*—it is true that *p* if and only if *p*—by means of any statement: "it is true that London has more than nine million inhabitants if and only if London has more than nine million inhabitants," "it is true that Boston is more beautiful than London if and only if Boston is more beautiful than London," "it is true that the British are nice people if and only if the British are nice people." The truth conditions of these statements are specified by what follows the phrase "if and only if," and they are *genuine* in the sense that they are *not* verification conditions: we can understand even statements that we cannot verify even in principle. Insofar as they are genuine, these truth conditions involve the world, involving the level of reality on which a certain statement focuses. If they are satisfied, the statement is true, and what follows the "if and only if" denotes a real situation to which the statement *corresponds*. As we know, for Putnam this is the only philosophically legitimate possibility for speaking of "correspondence": far from showing a single strong and metaphysically determined correspondence that binds *any* statement to the world, the "motley" of language shows a multiplicity of ways of corresponding to reality, as many ways as there are types of true statement.[17]

Such multiplicity reveals that the word "true" does not have an univocal use—the same use in all contexts in which it can be used with reason: just as in the case of words such as "object" and "exists," the word "true" is susceptible to a plurivocal use that is as sensitive to contexts as to the standards of evaluation therein. For example, the diversity of the content of the three statements above—a diversity given, to begin with, by their being about different domains of reality—results in three different uses of "true" governed by different standards for evaluating the correctness of those statements: a correspondence to the empirical state of affairs represented by the actual number of London's inhabitants at the time of the utterance of the first statement, a set of aesthetic criteria that can be considered valid and objectively recognized at the time of the utterance of the second statement, and a set of historical-ethical-social criteria that are valid and objectively recognized as such at the time of the utterance of the third statement. Note that, while the criteria for the correctness of the first statement are mostly set by the natural world, those of the other two reflect values established by a certain collectivity of human beings: what counts as true must therefore go through a certain amount of rational discussion in these cases, which shows how the notion of truth here matches the epistemic notion of warranted assertibility. In Chapter 8 we will have occasion to take up this aspect in relation to ethical statements. Note also that according to Putnam the notion of truth takes shape through norms for evaluating the correctness of statements, and is thus closely intertwined with the notion of *normativity.*

Finally, it should be noted that the plurivocal use of the word "true" is capable of making this word potentially *ambiguous* from a semantic point of view, since the multiple uses to which it can be adapted can give rise to the belief that there is nothing "common" to these uses themselves, and that the word, fluctuating between uses, changes meaning from time to time. It could therefore be thought that "true" is not a semantically and logically stable word, and that it is as many different homonymous words as there are contexts in which it is used, as is the case with words identical in spelling and pronunciation but having different meanings: *homonyms*, precisely, such as "bank," "step," "coach," etc., which in different contexts have different meanings arising from different etymological roots or from uses that may be from use to use literal or metaphorical, so that we learn those meanings separately. And this possible semantic ambiguity of "true," if real, cannot but affect the notion of truth as well, which would be similarly ambiguous and lacking its own *unity.* This is a danger of which Putnam is aware, since he points out how

the difficulty in giving a picture of our notion of truth [. . .] is to do justice simultaneously both to the *unity* of the notion, and the *plurality* of the correctness-
conditions that go with it and give it content (Putnam 2013a: 98).

One way to bypass this difficulty consists for Putnam in finding the desired
unity in the *property of equivalence*, the *property of normativity* and the *property of reference dependence*: these are constant properties of truth, that is,
present in each use of the word "true." The first[18] is the one shown by each
instance of the equivalence thesis and which Putnam links to Gottlob Frege:
for the great German mathematician and logician was among the first to
notice the semantic equivalence between a statement *p* and the statement by
which truth is attributed to *p*: "one can, indeed, say: 'The thought, that 5 is a
prime number, is true.' But closer examination shows that nothing more has
been said than in the simple sentence, '5 is a prime number'" (Frege 1892:
64); and again:

it is also worth noticing that the sentence "I smell the scent of violets" has just
the same content as the sentence "It is true that I smell the scent of violets."
So it seems, then, that nothing is added to the thought by my ascribing to it the
property of truth (Frege 1918–19: 354).

The property of equivalence is a constant of truth because it has a logical
character; indeed, it is "the chief logical principle governing the use of the
word 'true'" (Putnam 1994e: 50). Putnam highlights this by arguing that

when we extend the truth concept to a new area of discourse, we always take
care to preserve the logical properties, in particular disquotation, that appertain
to "true." In this respect "true" belongs to the family of the logical words (for
example, the connectives and quantifiers), which also are used in every area of
discourse (Putnam 2015d: 559–560).

Acknowledging the importance and centrality of the equivalence property,
however, is not the same as advocating the so-called *alethic deflationism*,[19]
namely the position subscribed to, albeit in different ways, by philosophers
such as Ramsey, Quine, and Horwich, and according to which the entire
analysis of the concept of truth can be reduced to a mere analysis of the use of
the word "true," in particular to a consideration of the instances of the equivalence thesis, these being what the concept itself comes down to, and nothing
more. Horwich, for example, believes that the application of the word "true"
to a proposition does not amount to attributing to the latter a special property
in terms of which to interpret the content of the concept, thus considering it
to be endowed with a metaphysically determined substance. For Horwich,

the word "true" has only an expressive utility, and neither it nor the concept governing its use can be used for explanatory purposes: "truth is metaphysically trivial" (Horwich 1998: 146). For Putnam, by contrast, what makes truth an important and by no means trivial metaphysical concept is both the second property, that of *normativity*—given by the fact that, as we have seen, every attribution of truth is governed by appropriate norms of correctness—and the property of *reference-dependence*, which concerns in particular the truth of descriptive statements—the empirical statements of ordinary language or the language of the natural sciences—and which together with the previous one makes truth a "substantive" notion.

Considered not as mere sequences of purely syntactic graphic signs on paper or sound material propagated through the air but, in the wake of Wittgenstein, as elements susceptible to a kind of use which involves the world within a particular linguistic community (cf. Putnam 2015b: 98–99 and 110), descriptive statements are linked to the portion of reality they pertain to by means of the reference relation brought about through the use of such names and predicates as may appear within them. For, which statements in a given language are true is determined by the reference of names and predicates: it is a *property of truth* [. . .] *that the extension of "true" depends on the extension of "refers"* (Putnam 2015c: 36).[20] Thus, if truth depends on reference, if "linguistic reference grows out of [. . .] perceptual transactions between organisms and things in their environments" (Putnam 2015c: 40), and if these transactions are direct and without intermediaries, then Putnam's picture of truth appears to be a conception perfectly in harmony with his natural realism.

Summing up the foregoing, we can say that Putnam emphasizes the irreducible and unpredictable variety that has to do with truth: the wide variety of domains in which truths can be formulated, where the empirical domain is but one among many; the wide variety of contexts of use of linguistic expressions within individual domains, each governed by distinct norms of correctness; the wide variety of types of evaluation of the truth-value of statements within individual contexts.[21] What emerges is an *alethic pluralism*[22] that combines the idea of the existence of an extendable family of uses of the word "true" with the idea of *normativity*, that is, that "to regard an assertion or belief or thought as true or false *is* to regard it as right or wrong" (Putnam 1994e: 69), such that

> it is a property of the notion of truth that to call a statement of any kind—not only an empirical statement, but also a mathematical statement, a statement of logic (e.g., *such-and-such a schema is valid*), an ethical statement, etc., etc.— true is to say that it has the sort of correctness appropriate to the kind of statement it is (Putnam 2013a: 97–98),

where, again,

> what sort of rightness or wrongness is in question varies enormously with the
> *sort* of discourse. *Statement, true, refers, belief, assertion, thought, language*
> [. . .] have a plurality of uses, and new uses are constantly added as new forms
> of discourse come into existence (Putnam 1994e: 69).[23]

The connection of these forms of discourse to the world is ensured—especially for descriptive forms—by the reference of words and predicates within statements, where the referential relation represents a linguistic evolution of a previous *direct* perceptual relation to objects.

Trying always to be faithful to what is our actual cognitive situation, Putnam thus advocates a *non-epistemic* notion of truth—which is an integral part of a *simply non-epistemic realism*—motivated by the fact that we understand without difficulty statements that to a great extent transcend what it is in our power to know and are therefore unverifiable even in principle. It is a notion that affects several areas of our practice—particularly the practice of the natural scientist—but it does not exhaust the plural face of truth: indeed, there are several areas of our practice in which an *epistemic* notion has its home. Rather, the idea that "it would be a great mistake to suppose that truth can always transcend warranted assertibility under 'ideal' (or good enough) conditions" (Putnam 2002a: 107) is what helps us to stem the skeptic and eliminate the BIVs hypothesis, while showing that the terms and predicates we employ in our linguistic exchanges possess an authentic reference worthy of study, whether it is knowable or not. The BIV hypothesis collapses thanks to the new picture of mind, language use and understanding, which shows how we are already *referentially* connected to the world: if we were not, we could not even understand the BIV hypothesis (cf. Putnam 2002a: 107–108), and this is enough to show its self-refuting character.[24] It is worth elaborating on the remarks in this last paragraph.

BIV AGAIN

We are now able to appreciate how behind the argument Putnam developed against the BIV hypothesis there are some of the central features of his philosophy. Some of these are left implicit, while the causal constraint on reference and the disquotational property of reference are made explicit.

Admittedly, Putnam's argument may leave one with a strange feeling—does it *really* prove that we are real human beings? It appears as the ultimate and final rejoinder to the skeptic—overwhelming and beyond appeal—but at the same time it inevitably seems a losing battle. One would more naturally and spontaneously say that proving the radical skeptic wrong or, equivalently, that we and the whole world exist, goes beyond the limits of the human cognitive faculties. And indeed, the skeptical challenge has been a thorn in the philosopher's side for centuries.

Traditionally, two kinds of reaction have been elicited by the skeptical challenge: one consists in the attempt to concoct a *direct* argument to prove the falsity of the skeptic's thesis, the other in the attempt to show this ungroundedness by presenting a reason why the skeptic's thesis should not be a cause for concern—this latter would then be an *indirect* argument. The feeling that something like the former argument is urgently needed was expressed by Kant in the following words:

> [. . .] it always remains a scandal of philosophy and universal human reason that the existence of things outside us [. . .] should have to be assumed merely *on faith*, and that if it occurs to anyone to doubt it, we should be unable to answer him with a satisfactory proof (Kant 1998: 121, "Preface to the Second Edition").

On the other hand, Heidegger can be seen as representing the idea that the second type of argument is sufficient, when, against Kant, he proposed that the scandal lay elsewhere:

> The "scandal of philosophy" does not consist in the fact that this proof is still lacking up to now, but *in the fact that such proofs are expected and attempted again and again* (Heidegger 1996: 190, Part One, VI, 43, a).

I think that Heidegger's is the right attitude to take in the face of the radical skeptic. We humans have obvious limits, and some of these limits are cognitive: we cannot know whatever is to be known, and we can reasonably expect that a greater or smaller part of the world is destined to remain unknown to us. Among the things that lie beyond the limits of our cognitive faculties there is the proof (an overwhelming proof) of the actual existence of the world outside us. But should this be a real cause for concern? A reasonable stance to take here is to say that, despite the lack of knock-down proof against the skeptic (or, equivalently, knock-down proof of the existence of the external world), we can present a sufficiently good argument—based on a sufficiently good reason—in favor of humans' genuine ability to gain knowledge. Such an argument would allow us to bracket the skeptical challenge, so to speak, and set it aside.

Does this mean that the radical skeptic can be defeated? Well, the answer is in the negative if defeat were necessarily to follow a knock-down argument, since we have just seen that we cannot secure an argument of this kind, the dismissal of the skeptical challenge being our only option. But perhaps even a good argument of a no-knock-down kind would do. This is what one can gather from a somewhat neglected piece by Putnam.

In replying to an essay Crispin Wright wrote on Putnam's BIV argument, Putnam distinguishes between two sorts of skeptic:

> One sort of skeptic—a very uninteresting sort—may raise a skeptical doubt only so that, no matter what premises one may rely on in answering the doubt, he or she can respond, "and how do you know *that?*" Obviously, this sort of skepticism—call it infinitely regressive skepticism—is "unanswerable," but equally obviously the existence of infinitely regressive skepticism shows only that justification must end somewhere (Putnam 1994i: 284).

The other sort of skeptic, the one Putnam had in mind in devising his argument against the BIV, moves from shared assumptions—that is, assumptions that both skeptic and non-skeptic hold—and aims to convince his or her opponent that all, or large part, of our empirical beliefs cannot amount to knowledge. Since this skeptic works from within our conceptual scheme, his or her stance can be termed "internal skepticism."

Now, if we buy this distinction, we realize that the former sort of skepticism is uninteresting because it appears to be similar to kids-questioning—that is, the kind of "Why?"questions that children often ask for no real reason, without actually expecting an answer, but only to strengthen their affective relationship with adults—and the latter sort of skepticism is worth responding to because it prompts us to reflect on the tenability of some of our basic assumptions. It is this latter sort of skeptic that we may hope to defeat—in the weak sense of "defeat" highlighted above—where Putnam's argument against the BIV can fulfil this hope. And in a full sense.

Indeed, it would be wrong to react to it in the following way:

> [. . .] the argument does not prove that I am not a brain in a vat; it makes the weaker point that if I were a brain in a vat I could not coherently think [or say] that I were. But from this it does not follow that the skeptical scenario that I might be a brain in a vat could not be true. This would only follow if the brain-in-a-vat scenario violated a physical law or contained a logical contradiction (Bernecker 2016: 58–59).

I think that two hidden assumptions are detectable behind this line of reasoning: one is that you should have a *direct* argument against skepticism, and the other is that a direct argument against skepticism is doomed to be *question-begging*. I think so because it seems to me that you can claim that Putnam's proof makes just the mentioned weak point if deep down you are convinced, on the one hand, that in order to make the stronger point it is necessary to prove beyond doubt the falsity of the BIV hypothesis and the truth of the hypothesis of the actual existence of the external world—that is, it is necessary to provide a direct proof—and on the other hand, that any such direct proof cannot but "assume" that human beings make a referential use of the terms of their language—which is exactly what was in need of proof.

The last point brings us to the recurrent misgiving regarding Putnam's proof that I hinted at at the end of § 6.2.6 in mentioning a possible *petitio principii*. Some authors—notably Anthony Brueckner (1986)—have based their rejection of Putnam's proof on its alleged question-begging character: "If it is agreed that you couldn't refer to BIVs if you were in the VAT scenario, don't you have to know that you are *not* in the VAT scenario before you can know that you can refer to BIVs—and thus know exactly the thing that the VAT argument is supposed to prove?" (Thorpe & Wright 2022: 66) Before commenting on this, let's look at another kind of misgiving. It is the following one:

> One may suspect that there is something wrong with this argument already when Putnam claims that the BIV hypothesis, although *physically* possible, may be ruled out by philosophy, through an *a priori* reasoning. For it is not clear how something can be physically (i.e., *a posteriori*) possible if it is already *a priori* impossible (Alai 1994: 12; cf. also Knowles 2023: 119–120).

It is in responding to this misgiving that we will be able to set ourselves on the right path to understanding the significance of Putnam's proof. When he first presented his argument, Putnam made it clear that he was trying to investigate

> the *preconditions* for *thinking about, representing, referring to*, etc. [. . .] by *reasoning a priori*. Not in the old "absolute" sense (since we don't claim that magical theories of reference are *a priori* wrong), but in the sense of inquiring into what is *reasonably* possible *assuming* certain general premises, or making certain very broad theoretical assumptions. Such a procedure is neither 'empirical' nor quite "a priori," but has elements of both ways of investigating (Putnam 1981: 16).

This is a revealing statement in that it shows, on the one hand, that his proof is not based on a traditional kind of a priori reasoning—that is, one performed in the armchair, so to speak—and on the other hand, that Putnam invites us to reflect on our epistemological situation—on our *epistemological standards*, which according to Putnam are not fixed once and for all. They are fallible and revisable, even eliminable, but as long as they remain in the field they possess enough strength to do their job. In brief, they exemplify the notion of "quasi-necessity relative to a conceptual scheme" we saw in Chapter 1: we cannot *currently* make sense of their falsity because they are a sort of Wittgensteinian *hinges*.[25] Therefore, they shape our ways of pursuing inquiries in general: "In our inquiries we can do no better than to start in the middle, relying on already established beliefs and inferences, and applying our best methods for re-evaluating particular beliefs and inferences and arriving at new ones" (Ebbs 2016b: 31). And among the established beliefs with which we start "in the middle" there are *substantive empirical beliefs* such as the beliefs that we are not always BIV, that there should be causal constraints on reference, and that magical theories of reference are wrong. These may sound like the usual empirical beliefs that can easily be false, but in actual fact they are *synthetic a priori*. If this is the case, then "I cannot express or understand the statement that I am always a brain in a vat without presupposing substantive beliefs and principles from which I may infer that I am not always a brain in a vat. I therefore conclude that the statement is self-undermining" (Ebbs 2016b: 35). Which is precisely Putnam's conclusion.

> Sometimes a thesis is called "self-refuting" if it is *the supposition that the thesis is entertained* or *enunciated* that implies its falsity. For example, "I do not exist" is self-refuting if thought by *me* (for any "*me*"). So one can be certain that one oneself exists, if one thinks about it (as Descartes argued).
>
> What I shall show is that the supposition that we are brains in a vat has just this property. If we can consider whether it is true or false, then it is not true (I shall show). Hence it is not true (Putnam 1981: 7–8).

It is interesting to point out the connection that the late Putnam glimpsed between his argument against the BIV hypothesis and some of Wittgenstein's reflections. This connection comes to light if we realize that, just as we cannot make sense of the "falsity" of our current epistemological standards due to their status as *quasi-necessary truths*, similarly we cannot make sense of the "truth" of the BIV hypothesis. The reason is that "to think that [BIV] are a possibility, I would have to assume that my entire view of the world is wrong in a way I am utterly unable to make cohere with anything I believe" (Putnam 2012h: 487). This is a point he finds in the discussion reported in *Lectures and Conversations on Aesthetics, Psychology, and Religious Belief*, where Wittgenstein maintains that the problem regarding the understanding of a claim such as "I believe in a Last Judgement" made by a believer is not a *linguistic one*—it is not that the non-believer does not grasp the meaning of the words in the utterance in question. He or she does grasp their meaning. Rather, the fact is that, despite the words' clear meaning, the utterance appears *unintelligible* to the non-believer (Wittgenstein himself) owing to the contrast between the utterance and (what the late Wittgenstein would call) the hinge propositions belonging to the non-believer's worldview. Here is Wittgenstein:

> If you ask me whether or not I believe in a Judgement Day, in the sense in which religious people have belief in it, I wouldn't say: "No. I don't believe there will be such a thing." It would seem to me utterly crazy to say this. [. . .]
>
> In one sense, I understand all he says—the English words "God", "separate," etc. I understand. I could say: "I don't believe in this," and this would be true, meaning I haven't got these thoughts or anything that hangs together with them. But not that I could contradict the thing.
>
> You might say: "Well, if you can't contradict him, that means you don't understand him. If you did understand him, then you might." That again is Greek to me. My normal technique of language leaves me. I don't know whether to say they understand one another or not (Wittgenstein 1966: 55).

In the same vein, according to Putnam we cannot really make sense of a hypothesis such as that of the BIV because we "haven't got these thoughts or anything that hangs together with them" and therefore the hypothesis itself appears unintelligible.[26] It is thus the appeal to this Wittgensteinian feature that adds to the argument of the 1980s based on the concept of meaning and, together with this argument, decrees the undoing of the BIV hypothesis—although it should not be forgotten that this feature is already implicitly contained in the notion of "quasi-necessity" developed by Putnam in the 1960s.[27]

To conclude, if we look at the BIV hypothesis from the perspective of our actual practices of inquiry and if we are convinced that to seek to answer the skeptic with a direct proof is to seek the impossible, we may realize not only that criticizing Putnam's proof as being question-begging is beside the point, but also that his proof is a neat way of silencing what Putnam calls the internal skeptic.

As I said at the end of the previous section, the BIV hypothesis collapses thanks to the new picture of mind, language-use and -understanding that Putnam arrived at in the last two decades of his constant reflection on these issues, and that shows how we are already *referentially* connected to the world. However, it is interesting to note that the path was to some degree clear decades before, as the following quotation from a 1975 piece of his shows:

> As language develops, the causal and noncausal links between bits of language and aspects of the world become more complex and more various. To look for any one uniform link between word or thought and the object of word or thought

is to look for the occult; but to see our evolving and expanding notion of reference as just a proliferating family is to miss the essence of the relation between
language and reality. The essence of the relation is that language and thought do
asymptotically correspond to reality, to some extent at least. A theory of reference is a theory of the correspondence in question.

Peirce and his Positivist and Therapeutic successors thought that good philosophy of language could clear up all the traditional problems of philosophy. I
would be repeating their mistake from a different standpoint if I claimed that the
realistic account of reference was the Philosopher's Stone (or even the Universal
Solvent). I do not claim this. But I do claim that it makes sense, and that the truly
best therapy is a sensible theory of the world (Putnam 1975c: 290).

NOTES

1. This is particularly true for statements for which there is currently no sure
method of specifying their truth-value, but which nevertheless have a determined
truth-value—or so it is believed.

2. This seems to echo Michael Williams's criticism of epistemic accounts of
truth in terms of ideal verification—accounts that he calls "neo-Peircean" and under
which Putnam's internal realism falls. Here is Williams: "For what are we supposed
to understand by ideal justification, verification, assertibility, etc.? Since the neo-
Peircean explication of truth is designed to decouple truth from anything we might
be inclined to cite as relevant to justification, thus doing justice to one of the realist's
central intuitions, we can give no further explication of this key notion in epistemic
terms. As soon as we try to be specific about what ideal justification involves, we
will be able to say of any belief that, even if it meets those standards, it might still
not be true, which is to say not ideally justified. So to the extent that we understand
anything by 'ideal justification,' we do so in virtue of a prior grasp of the notion of
truth. [. . .] We cannot give an informative analysis of truth in terms of ideal verification because that notion is either empty or understood in terms of truth" (Williams
1993: 196–197).

3. Criticism of the supposed intermediaries present in perception is a cornerstone
of McDowell (1994).

4. For an analysis of the perceptual aspect in Putnam's new form of realism, cf.
Macarthur (2004).

5. It should be noted that, in his later years, Putnam has also focused on the
question of whether the direct perceptions we have of the world are always "conceptualized." In his writings in the early 2000s Putnam opted for the idea of the con-
ceptualization of perception, along the lines of William James and John McDowell,
and then changed his mind decisively: "At the present time [2011], Hilla Jacobson
and I are working on a book on perception, and our view is that [. . .] not all experiences are conceptualized (contrary to McDowell) [. . .] the phenomenon James
called 'fusion,' in which the phenomenal character of experience 'fuses' with the

accompanying apperception, is not a feature of all experiences" (Putnam 2005a: 572). In this regard, cf. Jacobson & Putnam (2015) and the last three chapters in Putnam (2016a). For an illuminating review of the analytic philosophy of perception, cf. R. A. Putnam (1998b). In Travis (2005), Charles Travis supports the thesis that the idea of perception as something direct and unconceptualized has run consistently through the center of Putnam's philosophy, from at least 1960 to his last writings.

6. Cf.: "it is only in the context of such a practice of *talking about cats* that such a sentence as 'Tabitha won't drink milk' constitutes a statement. [. . .] *talking about things* is a vast and ever-expanding motley of *world-involving practices*" (Putnam 2004b: 65).

7. Dummett sharply criticized Putnam's natural realism (cf. Dummett 2007: 180 ff). For a defense of Putnam against this criticism, cf. Button (2013): 84–86.

8. For the verificationist treatment offered by Dummett of statements concerning the past, as well as his bold afterthoughts, cf. Dummett (1969, 1998, 2004).

9. "A simple example: a dog may be able to see a sign that says STOP, but even if it is taught to 'discriminate' that sign as pigeons are taught to discriminate various objects and various groups of objects, it will not experience the sign as *saying*, 'Stop!'" (Putnam 1994e: 192)

10. This example is analogous to Theorem 4 that Frederic Brenton Fitch demonstrates in the course of his illustration of the so-called "knowability paradox"—although Putnam claims that he was not familiar with Fitch's work at the time. According to the theorem, "For each agent who is not omniscient, there is a true proposition which that agent cannot know" (Fitch 1963: 138).

11. Here is how Putnam summarizes the point: "the impossibility of verifying the conjunction $P \wedge N \wedge S$ if it is true, where S is the statement that an observer cannot verify whether there is intelligent life in a region of space-time that the observer is unable to receive causal signals from and P is the possible empirical theory that tells us that (1) causal signals do not travel faster than light, (2) it is physically possible (and highly probable) that there are intelligent extraterrestrials, but (3) it is also physically possible that there are not, and (4) there are large regions of space-time that any particular physically possible observer is unable to receive causal signals from. $P \wedge N \wedge S$ is a statement that it is *logically* impossible to verify if it is true, and yet it is a statement that P itself tells us could be true (and even assigns a probability to)" (Putnam 2012b: 100–101).

12. It is also because of this (but not only because of this) that Putnam believes it is important to reason about statements such as K. In his words, "At present such conjectures are neither 'hypotheses' that we are testing nor 'linguistic memoranda to serve as a direct stimulus of further action.' They are statements that describe possible states of affairs that it is of intellectual value to think about, no more and no less than that" (Putnam 2015f: 796).

13. For a distinction between a narrow and a broad sense of "experience," see Putnam (2005a); on the different levels of form presented by reality, see also Putnam (2016c).

14. This is a point on which Ruth Anna Putnam's influence has been increasingly decisive, as we shall see in Chapter 8.

15. This explains Putnam's aversion to Quine's well-known criterion of ontological commitment, that to be is to be the value of a bound variable (cf. Putnam 1994e: 179; Quine 1948). For a discussion of the thesis of the univocity and the thesis of the plurivocity of the terms "being" and "existence" linking the contemporary debate to Aristotle's thought, see Berti 2001.

16. "My long engagement (and at times a struggle) with Wittgenstein's difficult and deep philosophical texts has indeed deepened my conviction that we need the sort of philosophical vision that is capable of recognizing the plurality of what Wittgenstein calls our 'language games' and of the 'forms of life' with which they are interwoven: the sort that is thoroughly anti-essentialist and free both of fantasies of reducing all the kinds of truth and objectivity there are to the kind of truth and objectivity characteristic of the exact sciences and of metaphysical fantasies with no connection either to real life or to real science" (Putnam 2015d: 547).

17. It is worth reiterating Putnam's extreme caution in speaking of correspondence. Although, as explained, there is a legitimate use of the notion, on several occasions he shows that he thinks it preferable to avoid mentioning it: "I fear that 'correspond' tends to suggest that truth depends on a relation (moreover, one and the same relation) between whole statements and something (reality? parts of reality? what sort of parts?) no matter what sorts of statements we are calling 'true'" (Putnam 2013a: 98–99). Strictly speaking, the only cases in which Putnam admits talk of correspondence are statements about the natural world that are true ("true empirical descriptions do correspond to real states of affairs") and true mathematical statements ("I think of mathematical truths as corresponding precisely to *possibilities and impossibilities* and relations between them [. . .] distinguishing between sentences which correspond to possibilities and impossibilities (and their relatives) and sentences which correspond to states of affairs fits well with both my account of empirical science and my modal-structuralist account of mathematics" (Putnam 2015d: 558–559). Many other truths, for example those of ethics such as "it is wrong to covet your neighbor's property," do not have the function of describing "a state of affairs in the natural world [or] a mathematical possibility or impossibility" (Putnam 2015d: 559), and so—strictly speaking—talk of correspondence is misplaced here.

18. In several places Putnam calls it the "disquotation property" (an expression introduced by Quine), referring to the metalinguistic version of the Tarski-style equivalence thesis: indeed, Tarski's well-known example is "'Snow is white' is a true sentence if and only if snow is white." In this version, the word "true" produces the effect of self-eliminating and deleting the quotation marks, as seen to the right of the "if and only if." In his later writings, however, in order to avoid being assimilated among the alethic deflationists (whose position he has consistently opposed), Putnam prefers not to employ the word "disquotation" to refer to this property of truth (see Putnam 2015b: 94).

19. Frege, in fact, was not a deflationist with regard to truth, and neither was Wittgenstein, according to Putnam's interpretation (others, on the contrary, make Wittgenstein one of the main models of an alethic deflationist: see Horwich 2016).

20. It is precisely by leveraging the notion of reference that Putnam builds his fundamental objection to alethic deflationism, once he points out that the equivalence thesis presupposes the notion of translation: "That the notion of translation is needed for disquotation and therefore needed by deflationists (since their thesis *is* that grasp of disquotation is all that is needed for an understanding of truth) is widely recognized. But what I have not seen discussed by deflationists, let alone taken seriously, is the thought that *translating sentences presupposes knowing what their descriptive constituents refer to*" (Putnam 2015c: 38).

21. Cf.: "We make up uses of words—many, many different uses of words—and the senses of 'agree' in which our various sentences 'agree' with reality, when they do, are plural indeed" (Putnam 1994g: 302).

22. In the current literature on truth, alethic pluralism constitutes a constellation of non-homogeneous positions, ranging from the idea that there are multiple different properties of truth to the idea that there are multiple properties that play the role of truth (roughly, the former is expressed in Wright 2001 and the latter in Lynch 2009). For a general illustration see Pedersen and Wright (2013) and (2016), Lynch, Wyatt, Kim, & Kellen (2021), Part VII.

23. For a detailed analysis of the many dimensions of the normativity of truth and their connection to alethic pluralism, cf. Ferrari (2022).

24. I think it is the support of metaphysical realism in the form of a *simply non-epistemic realism* that motivates the following statement: "I no longer accept the conclusion of [*Reason, Truth and History*], which was that the Brain in a Vat argument refutes metaphysical realism" (Putnam 2016a: 222). Likewise, Joshua Rowan Thorpe recognizes that a kind of metaphysical realist (the simply non-epistemic one, in my terminology; the external realist, in Button's terminology, which Thorpe follows) can happily embrace Putnam's argument against the BIV hypothesis: "the external realist [. . .] can be rid of [. . .] scepticism, whilst continuing to claim that the world is in some very strong sense mind independent. [. . .] The external realist will breathe a sigh of relief, for she endorses a picture of the world on which we easily might have been [in such a sceptical scenario]" (Thorpe 2018b: 393).

25. The notion of a "hinge" as introduced by Wittgenstein (1969) is discussed and extended in Coliva (2015).

26. Incidentally, it is to the idea of unintelligibility as emphasized by the late Putnam that we can trace the fundamental reason why the Putnam of the Eighties claimed that, if we came across "a vast number of people [who firmly think they are BIV and therefore] have a self-contained belief system which violently disagrees with ours" (Putnam 1981: 131), then we would judge them "crazy in the sense of having *sick* minds" (Putnam 1981: 132; see Chapter 8, Section "Scientific Language and Values").

27. Indeed, as already mentioned, a quasi-necessary statement is analogous to a Wittgensteinian hinge. In this regard, Luigi Perissinotto keenly observed (in conversation) that, in the end, the appeal to the issue of unintelligibility makes Putnam's proof of the 1980s redundant.

Chapter Eight

Moral Philosophy and Philosophy of Religion

WHY PHILOSOPHY

Reflections on the usefulness and significance of philosophy are not uncommon in Putnam's writings. These are the aspects in which the influence of the pragmatist tradition on his thinking is most noticeable, as we will appreciate in Chapter 9. In *Pragmatism: An Open Question* he states, for example, that:

> We want ideals and we want a world view, and we want our ideals and our world view to support one another. Philosophy which is all argument feeds no real hunger; while philosophy which is all vision feeds a real hunger, but it feeds it Pablum (Putnam 1992a: 23).

Along the same lines, a few years earlier, Putnam declared that the philosopher has a double task:

> To integrate our various views of our world and ourselves [. . .], and to help us find a meaningful orientation in life. Finding a meaningful orientation in life is not, I think, a matter of finding a set of doctrines to live by, although it certainly includes having views; it is much more a matter of developing a *sensibility* (Putnam 1989a: 52).[1]

There is a great consonance here with what two great contemporary philosophers—both greatly admired by Putnam—say on this subject: Wilfrid Sellars and Stanley Cavell. According to the former, "the aim of philosophy, abstractly formulated, is to understand how things in the broadest possible sense of the term hang together in the broadest possible sense of the term" (Sellars 1962: 37). Sellars' dictum is quoted approvingly in several places in

Putnam's writings, although Putnam does not believe that it succeeds in fully identifying the purpose of philosophy.

> Speculating about how things hang together requires knowledge, sophistication, imagination, the ability to draw out conceptual distinctions and connections, and the ability to argue. The activity appeals to our curiosity, our daring, our delight in intellectual keenness. But speculative views, however interesting, or well supported by argument, or insightful, are not all we need (Putnam 1989a: 52).

That is why he turns to Cavell. Cavell argued that the importance of philosophy lies in its being "the education of grownups" (Cavell 1979: 125): in helping to identify ways to answer the various questions that inevitably dot our existence and that have much in common with the "whys" raised by children. In both cases, the "orientation" aspect that philosophical reflection must have, if it is to be of any use, stands out. However, it is by placing himself in the wake of one of the fathers of pragmatism, John Dewey, that Putnam shows the other important feature that philosophy cannot in his view fail to possess and which came up in the second quotation above: that of shaping our sensibilities. The most plausible way to characterize philosophy, Putnam argues, is to understand it as a *reflective transcendence*, which is to say:

> standing back from conventional opinion, on the one hand, and the authority of revelation (i.e., of literally and uncritically accepted religious texts or myths) on the other, and asking "Why?" Philosophy [. . .] thus combines two aspirations: the aspiration to justice, and the aspiration to critical thinking (Putnam 2004a: 92).

Thus, it is these two aspirations that shape our sensibility and make us "ethical" people, attentive to the needs and desires of others—both on an individual and on a social level. Aspirations undoubtedly linked to each other and endowed with equal importance, although in general it is the latter that ends up playing a prominent role by encompassing the former, exactly as Dewey intended. The "criticism of criticisms" is in fact Dewey's characterization of philosophy (see Dewey 1925: 398), meaning by this phrase

> not just the criticism of received ideas, but higher-level criticism, the "standing back" and criticizing even the ways in which we are accustomed to criticize ideas, the criticism of our ways of criticism (Putnam 2004a: 96).

All of this with the aim of improving our position in the world, that is, our relationship with our environment and with others—as a fundamental part of our environment—by making that relationship and the knowledge we have of it increasingly reliable and refined. This makes us appreciate how for Putnam (a thinker so versatile that he turns his attention to science, mathematics, logic, language, mind, knowledge, reality, truth, and religion) philosophy is ultimately about nothing more than the fundamental Socratic question that has become central to ethics, namely, "How to live?" (cf. Putnam's 1976b close) In this way, Putnam reconnects (through Pierre Hadot) with the ancient Greek tradition: the tradition in which

> the *central* ethical question was not "what are the right rules of conduct?" (although that was an important question), nor even "what are the several virtues?" but "what should be the supreme aim of a well-lived human life?" (Putnam 2005b: 159–160)

But if this is the question to which philosophy must find an answer—indeed, as many answers as the particular declinations that question may receive in the most varied circumstances of our lives—then it follows that the focus of philosophical attention *cannot* be directed to the identification of general norms susceptible "in the abstract" to be applied to any problematic situation; on the contrary, it is the "concrete" *practice* in which we find ourselves engaged on a daily basis that assumes central importance, the practice in which we are forced to deal with existential difficulties of varying depth and degree, and that compels us to make an enormous variety of moral judgments—which makes ethics "a motley *squared*" (Putnam 2004a: 72).

THE FUNCTION OF ETHICS

Indeed, the primary function of ethics is not "to arrive at 'universal principles,'" nor "to produce a 'system,' but to contribute to the solution of practical problems" (Putnam 2004a: 4). If indeed we want ethics to be a system, then we cannot help but see it as "a system of interrelated concerns [which are] as mutually supporting but also in partial tension" (Putnam 2004a: 22). These concerns are certainly about such things as democracy, tolerance, and pluralism, but, Putnam tells us, for the purpose of identifying the most fundamental concerns we need to look at what emerges from the thinking of philosophers as diverse as Aristotle, Immanuel Kant, and Emmanuel Levinas. The first because his ethics is centered on the question "what is it that makes a human life admirable?" and implicitly urges us to reflect on what makes an existence worthy of admiration and on the ways to develop the potential each of us has to live a full and satisfying life. The second because he has identified an indispensable notion, that of the "categorical imperative." But, notice, what makes the categorical imperative important in Putnam's eyes is not this imperative in itself—that is, understood as a practical guide ("as a guide it scarcely goes beyond the Golden Rule," Putnam 2004a: 25)—but rather the idea behind it. It is in fact to be understood

> as a powerful statement of the idea that ethics is *universal*, that, insofar as ethics is concerned with the alleviation of suffering, it is concerned with the alleviation of *everyone's* suffering, or if it is concerned with positive well-being, it is concerned with *everybody's* positive well-being (Putnam 2004a: 25).

It is the belief that inspires the categorical imperative that makes it important; it is not important in itself. Indeed as an absolute principle, applicable in any circumstance and independent of our judgment, it cannot be part of ethics as Putnam understands it.

Finally, in order to understand what concerns ethics is substantiated by one must turn to Levinas. For it is he who identified the irreducible foundation of ethics in "*my* immediate recognition, when confronted with a suffering fellow human being, that *I* have an obligation to do something" (Putnam 2004a: 24).

Thus, ethics is not a single set of principles, as if it were "a noble statue standing at the top of a single pillar" (Putnam 2004a: 28), but represents a deep, concrete, first-person commitment to others; not a statue on a single pillar but a table with many legs, as many as the number of concerns we feel we have.

> We all know that a table with many legs wobbles when the floor on which it stands is not even, but such a table is very hard to turn over, and that is how I see ethics: as a table with many legs, which wobbles a lot, but is very hard to turn over (Putnam 2004a: 28).

PRACTICE

This, then, elucidates the centrality to ethics of the *practice* in which we are immersed: because it is in the latter that the myriad of personal and interpersonal problems arise that are capable of arousing those particular concerns in which ethics consists and that prompt us to find appropriate solutions. To be able to address these problems is to take on genuine concerns about the welfare of others at any level. But—it should be emphasized—the solutions at issue here are "appropriate" in the sense of being tailored to the particular problem addressed: they are specific and specifically identified solutions, and not general and abstract solutions to be considered in principle adaptable to all problem cases. The problems we are talking about are indeed *practical* in the literal sense, that is, "'problems we encounter in practice,' specific and situated problems, as opposed to abstract, idealized, or theoretical problems" (Putnam 2004a: 28). And—one would ask—how could it be otherwise? How could the infinite variety of morally interesting situations be subsumed under a narrow group of abstract principles?[2]

It would seem that at the bottom of ethical reflection developed in the Western philosophical tradition there sometimes hovers the implicit belief that problems relevant to ethics admit of precise and well-defined solutions, even that they are amenable to solutions endowed with objective validity of an "absolute" character. Interestingly, the existence of such solutions would render moral conflicts entirely "illegitimate," that is, devoid of any plausible reason, and thus subsisting only because the factions involved fail to realize what is the only right position in the dispute. Consider, for example, a Kantian. Since for her

> the objectivity of moral judgments follows directly from the nature of principles, [we have] that, in any genuine moral dispute, at least one party must be wrong (Davidson 1995: 40).

In any sphere where firm and absolute laws dictate right conduct, any disagreement can only be brought about by, say, conceptual shortsightedness, intellectual laziness, or simple inattention, and would therefore be illegitimate—the result of an inability to grasp the well-defined solution that a certain principle establishes. The existence of such laws in ethics is very difficult to justify, and so is the strong kind of objectivity they would confer on our moral judgments. It is probably precisely the impossibility of achieving this kind of objectivity in ethics that has spread the belief that moral judgments are hopelessly subjective, as opposed to judgments about facts (considered as the model of objectivity par excellence), and that has led to the so-called *fact/value dichotomy*: the idea that judgments about facts and those involving values (ethical, aesthetic, legal, religious, etc.)[3] are of a radically different nature. Putnam has for decades tenaciously argued against this view, to the point of becoming one of the main points of reference for those who believe it is possible to argue for the objective validity of ethical discourse without presupposing absolute laws or principles, thus taking on a difficult challenge. As Donald Davidson acknowledges,

> it is harder to make a convincing case for the objectivity of moral judgments if, unlike the Kantian or the utilitarian, you hold that there are legitimate moral conflicts. I am convinced of the objectivity of many moral principles, but I believe such principles can come into genuine conflict; it can happen that we ought to perform some act and also ought to refrain from performing it (Davidson 1995: 41).

Davidson's statement would not be out of place in the mouth of Putnam, who for his part argues that practical problems, unlike the idealized thought experiments of the philosophers, are typically "messy." They do not have clear-cut solutions, but there are better and worse ways of approaching a given practical problem (Putnam 2004a: 28–29),

where any claim that a certain approach is better, or worse, than another is made with reference to a criterion of objective validity. Statements of this kind are pervasive in our daily practice, and this shows that in speaking, in discussing, in arguing we presuppose such a criterion:

> we make and cannot escape making value judgments of all kinds in connection with activities of every kind. Nor do we treat these judgments as matters of mere *taste*; we argue about them seriously, we try to *get them right*, and [. . .] we use the language of objectivity in our arguments and deliberations—for example, we use the same laws of logic when we reason about an ethical question that we use when we reason about a question in set theory, or in physics, or in history, or in any other area (Putnam 1994b: 154).

And, note, the *language of objectivity* applied across the board is just enough to dissolve the fact/value dichotomy. The question remains, then, whether such objectivity really exists: let us get there by examining the arguments on the basis of which Putnam believes that the dichotomy between fact and value judgments can only collapse.[4]

AN ARGUMENT AGAINST THE FACT/VALUE DICHOTOMY: THE NOTION OF FACT

Putnam first notes that, contrary to what one might think at first glance, the subject of fact and value is of interest to everyone.

> In this respect, it differs sharply from many philosophical questions. Most educated men and women do not feel it obligatory to have an opinion on the question whether there really is a real world or only appears to be one, for example. Questions in philosophy of language, epistemology, and even in metaphysics may appear to be questions which, however interesting, are somewhat optional from the point of view of most people's lives. But the question of fact and value is a forced choice question. Any reflective person *has* to have a real opinion upon it (Putnam 1981: 127).

Only, Putnam continues, the opinion underlying the general ethical attitude manifested by most people expresses precisely the aforementioned dichotomy between beliefs concerning facts and beliefs concerning values, so much so that the "view that there is no fact of the matter as to whether or not things are good or bad or better or worse, etc. has, in a sense, become *institutionalized*" (Putnam 1981: 128): that is, it is so entrenched that even the strongest philosophical argument that could be devised to show how such a dichotomy lacks the slightest rational basis—exactly as Putnam believes, the argument that has in addition the greatest persuasive force that a philosophical argument could ever possess—would scarcely be effective. Yet it is a rationally indefensible dichotomy; there are basically two paths Putnam takes to illustrate this. The first leads one to consider what a fact is, and shows how the dichotomy presupposes a "narrow" notion of it; the second way shows instead that there can be no description that is not at the same time an evaluation, and vice versa. Let us look at the first.

Putnam begins by noting how it has happened several times in our philosophical tradition that an innocent distinction between different notions is surreptitiously loaded with a metaphysical valence. One such example is offered by what is known as *Hume's law*, a name owing not to David Hume but to those who would later advocate a *marked* fact/value distinction—a dichotomy in fact. What we get from what the great empiricist states in the third book of the *Treatise of Human Nature* is simply that an *ought* cannot be derived "logically" from an *is*. Here is Hume:

> In every system of morality, which I have hitherto met with, I have always remark'd, that the author proceeds for some time in the ordinary way of reasoning, and establishes the being of a God, or makes observations concerning human affairs; when of a sudden I am surpriz'd to find, that instead of the usual copulations of propositions, *is*, and *is not*, I meet with no propositions that is not connected with an *ought*, or an *ought not*. This change is imperceptible; but is, however, of the last consequence. For as this *ought*, or *ought not*, expresses some new relation or affirmation, "tis necessary that [. . .] a reason should be given, for what seems altogether inconceivable, how this new relation can be a deduction from others, which are entirely different from it" (Hume 1739–40: III, Part 1, 1: 469).

Hume is making a wholly defensible observation about the unjustifiability (i.e., the lack of a "reason") of the relation of logical derivability between statements containing the verb *to be*—descriptive statements—and statements containing the modal verb *ought*—prescriptive statements, that is, expressing a norm. This relationship is unjustifiable because only descriptive statements can be logically deduced from descriptive statements, never prescriptive statements (and vice versa): thus, there is no way to go from *is* to *ought to be* (and vice versa). And, notice, Hume's remark concerns logic and logically permissible inferences, not metaphysics. As we have just seen, all that Hume literally states is that "in all the 'systems of morality' he had met, the author would start in 'the ordinary way of reasoning,' proving, say, the existence of God or describing human society, and suddenly switch from 'is' and 'is not' to 'ought' and 'ought not,' for example from 'God *is* our creator' to 'we *ought* to obey him.' No explanation was ever given of this 'new relation'" (Putnam 2002a: 149). There is no necessary metaphysical implication in such a claim. However, in subsequent philosophical reflection

> no one, including Hume himself, ever takes it as merely a claim about the validity of certain forms of inference, analogous to the claim "you cannot infer 'p & q' from 'p or q'" (Putnam 2002a: 14),

with the consequence that the innocuous claim that there is an obvious difference between judgments of fact and judgments of value (evidenced by their not being logically derivable from each other) has been charged with a *metaphysical* valence. As if to say that this is how things are "in reality." The question then arises as to why it was "consequential" for Hume, and those who followed him on this point, to associate a metaphysical coloring with this observation of a logical character. The answer is that they involved some further premise not made explicit in the reasoning above, bending it in a direction that leads them to regard the distinction between descriptive and prescriptive statements (or, more generally, between factual and value judgments) as a dichotomy. And indeed, as far as Hume is concerned, the responsibility for this undue metaphysical burden lies with his metaphysics of "matters of fact." What these are is quickly stated.

The whole sphere of human rationality, the sphere in which the tradition prior to the Scottish philosopher distinguished knowledge and belief—the former conceived as certain while the latter as probable—is reinterpreted by him on the basis of what has been called "Hume's fork": the distinction between *relations of ideas* and *matters of fact*. Statements expressing relations between ideas are a priori, since they are derived through the operation of thought alone and are therefore true independently of all experience. That the sum of the interior angles of a plane triangle is 180 degrees or that 2 plus 5 equals 7 are derived by intellectual and not empirical means: these are thus supremely certain truths—denying them leads to contradictions. In contrast, statements concerning matters of fact can be denied without this constituting an immediate contradiction: simply, what I believe now by virtue of a certain experience of facts in the world (say, that it was the wear and tear of time that caused a certain bridge to collapse) can be contradicted in the future by a better ascertainment of the facts. Note that for Hume the sphere of relations of ideas and that of matters of fact represent two exclusive and exhaustive categories: a statement belongs to one category or the other, and they exhaust the scope of rationality. The distinction between the two categories is sharp, dichotomous. And it is metaphysical in nature: *such* is human and non-human reality, according to Hume. And so is our rational activity, which can ultimately rely on only one source: that of experience. For ideas, according to Hume, are nothing more than "faded" sensory impressions, and their nature therefore derives from experience—which among other things gives us access to facts.

What is important to understand, however, for the purposes of the argument we were making is that the notion of fact that emerges from Hume's philosophy restores facts as something "experienceable" in a narrow sense, that is, as something—an *impression*—that can be perceived through one or more of our five senses. But "if *this* is the notion of a fact, then it is hardly surprising that ethical judgments turn out not to be 'factual'!" (Putnam 2002a: 22): no one has ever sensorially experienced an ethical, aesthetic, legal, etc., fact—and never can. It thus becomes evident how this notion of fact (typical of classical empiricism and surviving into the twentieth century thanks to the support of some logical positivists) is too *narrow*: it is the development of scientific research that would gradually make it clearer. Facts concerning entities that according to a rigidly empiricist view should have been considered unobservable and therefore non-existent were considered (after Hume and, ironically, at the height of the logical positivist movement) perfectly legitimate and part of what a scientist worthy of the name considered to be objects of investigation: facts about bacteria, atoms, subatomic particles, curved space-time, and others revealed that the "idea that a 'fact' is just a sensible 'impression' would hardly seem to be tenable any longer" (Putnam 2002a: 22).

The empiricist conception of fact was thus being "overthrown more by the progress of science itself than by philosophical argument" (Putnam 2003: 114–115), although it is up to the latter to explicate the meaning of such an overthrow by illuminating its various aspects. And Putnam does so masterfully, summarizing the intellectual history of logical positivism, heir to the Humean tradition.

After recalling that "the positivists found themselves pressed more and more to abandon their initial notion of a fact, which was somewhat similar to Hume's, in order to do justice to the revolutionary science of the first half of the twentieth century" (Putnam 2002a: 21), Putnam observes that, paradoxical as it may seem, it was precisely their gradual revision of that notion that destroyed the basis on which the dichotomy between fact and value had been erected. Indeed, if the notion of "fact" is to include also that which can be described by statements that contain terms referring to scientifically legitimate but unobservable entities (the so-called *unobservable* terms such as "charge," "electron," "gravitational field," "gene," and "bacterium"), then the logical positivists sought, first, to place the requirement that unobservable terms must be *reduced* to observable ones, and later, faced with the impossibility of making such reductions in the vast majority of cases, they stipulated that they should be considered as *primitive* (or *theoretical* versus *observational*) terms of a scientific theory that have meaning to the extent that *the theoretical system as a whole* (i.e., the theory reformulated in a formalized language) allows for successful predictions of experience. In short, predictive success of a theory containing theoretical terms would be an indirect sign of their empirical significance.

However,

> to *predict* anything means (to logical positivists) to *deduce observation sentences from a theory*. And to deduce something from a set of empirical postulates, we need not only those postulates but also the axioms of mathematics and logic (Putnam 2002a: 29),

axioms that obviously do not express facts and yet are indispensable in science. They are *analytic*, and acceptable for logical positivists in scientific language as such. "Thus the search for a satisfactory demarcation of the 'factual' became the search for a satisfactory way of drawing 'the analytic-synthetic distinction'" (Putnam 2002a: 29). And it is the philosophical argument here, rather than the aforementioned "progress of science," that shows that there can be nothing satisfactory in such a search.

Putnam, we now know, played a large role in this, taking up and refining the legacy of a philosopher of Quine's caliber. As we saw in Chapter 1, Quine is credited with one of the most decisive critiques of the analytic/synthetic dichotomy, a dichotomy linked to the fact/value dichotomy by the way that one horn of each dichotomy relies on the narrow notion of fact typical of empiricism. We pointed out how Putnam does not, however, follow Quine in what would seem to be the conclusion of his argument, namely, that there is *no* point in distinguishing a class of analytic truths from a class of truths subject to observational tests, and how he believes instead that "the notion of an analytic statement can be a modest and occasionally useful notion, but so domesticated that it ceases to be a powerful philosophical weapon that can perform such marvelous functions as explaining why mathematical truths pose no problem at all for empiricism" (Putnam 2002a: 13). Putnam thus believes that the class of analytic truths is not empty, albeit very narrow, and exactly as in the case of Hume's law he points out that the analytic/synthetic distinction, far from being a dichotomy, is a modest distinction, linguistic in character and lacking in metaphysical scope. But if it is devoid of metaphysical scope, then it does not help to clearly demarcate what is *factual* from what is not and is rather the result of *convention* (as the logical positivists understood mathematical and logical truths). On the contrary, "fact and convention interpenetrate without there ever being any sentences that are true by virtue of fact alone or true by virtue of convention alone" (Putnam 2002a: 138; see above Chapter 2, Section "Fact and Convention"). As Quine summarized the process in words that have rightly become famous:

> The lore of our fathers is a fabric of sentences. In our hands it develops and changes, through more or less arbitrary and deliberate revisions and additions of our own, more or less directly occasioned by the continuing stimulation of our sense organs. It is a pale gray lore, black with fact and white with convention. But I have found no substantial reasons for concluding that there are any quite black threads in it, or any white ones (Quine 1963: 125).

Thus, if there is no notion of fact independent of the notion of convention, it is not only the analytic/synthetic dichotomy that crumbles but also the fact/ value dichotomy. Taking up Quine's image, the economist and philosopher Vivian Walsh has argued that "if a theory may be black with fact and white with convention, it might well [. . .] be red with values" (Walsh 1987: 862), meaning that "the possibility cannot be excluded that *values* are also a factor that permeates science as a whole" (Putnam 2003: 114). Indeed, given that there is no "notion of *fact* that contrasts neatly and absolutely with the notion of 'value'" (Putnam 2002a: 26), on closer inspection, among the notions of fact, convention and value there is a *triple entanglement* (cf. Putnam and Walsh 2007: 150 and 168), an entanglement underlying our description of facts both in the language of science and in everyday language.[5]

AN ARGUMENT AGAINST THE FACT/VALUE DICHOTOMY: THE ENTANGLEMENT OF DESCRIPTION AND EVALUATION

Scientific Language and Values

In order to show how the triple entanglement is at work in both scientific and everyday language, Putnam first invites us to appreciate how scientific research itself—the research that the philosophical tradition has wanted to see as devoid of all value and solely devoted to the pure ascertainment of facts—presupposes a particular kind of values: *epistemic* values. To bring out such values in all their evidence, he proposes a kind of thought experiment.[6]

Suppose that there is a very large group of people (as many as live in Australia, for example) who unshakably believe that *all* human beings on Earth are brains in a vat—not flesh-and-blood creatures, as we know, but brains that for some mysterious reason are kept alive inside a vat filled with nutritive fluids, unable to realize that they are in this deplorable situation. And that they are unable to realize it because the brains (*our* brains, according to them) are connected to a super modern computer implemented with a myriad of different programs that send complex sensory inputs to the brains giving them the impression that they are experiencing the many situations of a "normal" life. *Our* belief system and *theirs*—that of the Australians so deeply convinced of such a miserable existential condition, and convinced, say, because they are followers of a certain Guru (it is still Putnam who is imagining it)—then disagree radically.

Now, although this is not a disagreement of a *scientific* kind (we can imagine that these Australians share with us the same discoveries and progress in science, probably differing from us only in some scientific predictions), nor even of an *ethical* kind (they may have a moral system very similar to ours), what, according to Putnam, we would say about these hypothetical Australians is that their worldview is *crazy*: they "would be regarded crazy in the sense of having *sick* minds" (Putnam 1981: 132). Why?

A first reason is that their cognitive system is *incoherent* in a particular sense of the term, the sense according to which it fails to provide the slightest explanation of the correctness of the system as a whole and the truth of the statements that constitute it. Indeed, in general, being able to justify internally the cognitive claims made, to show how one's perceptual and non-perceptual knowledge is reliable, and thus to show how the statements expressing that knowledge are true, constitutes an important constraint on our system itself: if the system of concepts, beliefs, and knowledge we have developed does not satisfy such a constraint, that is reason enough to amend it in appropriate ways. This kind of coherence then represents a methodological virtue that the system of the imaginary Australians lacks, since they "have themselves postulated an illusion so perfect that there is no rational way in which the Guru of Sydney can possibly *know* that the belief system which he has adopted and persuaded all the others to adopt is correct" (Putnam 1981: 133).

Another reason is that their system is not *inclusive*, in the sense that it fails to determine whether the laws of nature that "seem" to be valid to all creatures who believe they are brains in a vat are actually the laws valid outside the vat. In other words, the system in question also lacks the methodological virtue of comprehensiveness, "for it does not, even in its own terms, tell what the true and ultimate laws of nature are" (Putnam 1981: 133). In addition to this, it is in conflict with the maxim known as "Ockham's razor," the maxim that requires us to consider as existing only those entities that it is strictly necessary to: if we therefore consider a theory that obeys Ockham's razor to be *functionally simple*, we can say that the Australians' system of thought does not possess such virtue since it "postulates all kinds of objects outside the vat which play no role in the explanation of our experiences" (Putnam 1981: 133).

Now, the purpose of this thought experiment is to bring to light the set of values presupposed by a conceptual system in general and by science in particular, that set which begins to delineate its contours in the criticism we move against worldviews we consider inadequate. In the course of that criticism we can thus realize that underlying the relentless effort to construct and refine the total conceptual system by which we provide explanations and predictions of reality is an intrinsic desire to construct a representation of the world that is coherent, comprehensive, and functionally simple, to which we can add the values constituted by instrumental effectiveness, plausibility, reasonableness, simplicity, naturalness, preservation of past doctrines, and even beauty: the "inner perfection" of a theory, as Einstein puts it.[7] These are all *values* (since each represents a normative parameter) of an *epistemic* nature: they indicate "'what ought to be' in the case of reasoning," or "'the admirable' in the way of (scientific) conduct" (Putnam 2002a: 31 and 135), as the classical pragmatists and Peirce in particular argued.

The epistemic values that in Putnam's thought experiment we saw guide us in the criticism of systems of thought alternative to our own are exactly the criteria that guide us in evaluating statements and theories—when it comes to accepting or discarding them—or choosing between competing theories. Contrary to the view held by much philosophy of science, and in particular by logical positivists and Popperian neo-rationalists, a scientist's rational activity is not solely in the comparison of statements and theories with observational data, a comparison governed by strict formal rules of confirmation or falsification: rather, the scientific method is an informal matter of trade-off and balancing between empirical data and theories, Putnam points out, and very often the choice between alternative theories intended to explain the same facts is not based on observational data at all. Case in point:

> both Einstein's theory of gravitation and Alfred North Whitehead's 1922 theory [. . .] agreed with special relativity, and both predicted the familiar phenomena of the deflection of light by gravitation, the non-Newtonian character of the orbit of Mercury, the exact orbit of the moon, among other things. Yet Einstein's theory was accepted and Whitehead's theory was rejected fifty years before anyone thought of an observation that would decide between the two (Putnam 2002a: 142),

where the "judgment that scientists explicitly or implicitly made, that Whitehead's theory was too 'implausible' or too 'ad hoc' to be taken seriously, was clearly a *value judgment*" (Putnam 2004a: 68) based on the epistemic values referred to above.

The statement with which Putnam summarizes all this is very indicative: given that the only way to get a picture of the world is to go through the

representations of it that we elaborate and evaluate in the course of our incessant rational activity, and given that we do not "have some way of telling that
we have arrived at the truth *apart from* our epistemic values" (Putnam 2002a:
32), we come to realize that "the 'real world' depends upon our values"
(Putnam 1981: 135), since without them "we have no world and no 'facts'"
(Putnam 1981: 136). As if to say that physics without values (epistemic, ethical, aesthetic, etc.) is blind.

Ordinary Language and Values

Scientific facts and epistemic values are thus closely intertwined. But it is
our language in its generality that manifests an even deeper entanglement
between facts and ethical, aesthetic, legal, political, religious, and any other
kind of values, an entanglement that breaks down the dichotomy between
descriptive and evaluative uses and that stands out especially in the case of
some particular adjectives.

These include the adjective "cruel": this is an adjective that can be used
both for evaluative purposes (when we say things like "He behaved cruelly
to her") and for descriptive purposes (as when a historian says "He was a
very cruel monarch"). In metaethics, the concepts that preside over the use of
adjectives of this kind are called *thick* ethical concepts, to distinguish them
from the *thin* ethical concepts underlying the use of adjectives and expressions such as "good," "right," "ought," and the like, which are solely evaluative and normative in nature. Other adjectives that belong to this class are
"brave," "wise," "compassionate," "resourceful," "considerate," "jealous,"
plus a great many others and their antonyms:

> the judgment that someone is inconsiderate may indeed be used to blame; but
> it may be used simply to describe, and it may also be used to explain and to
> predict. [. . .] And similarly, "jealous" may be a term of blame and may be used
> without any intention of blaming at all. (Sometimes one has a perfect right to be
> jealous) (Putnam 1981: 138–139).

A very heated debate has been alive around thick concepts for decades,
with differing positions, including those who believe that they are after all
factual concepts, those who believe that they should be broken down and
analyzed into two components, the descriptive and the evaluative, and those
who believe that such a breakdown cannot be performed.[8] Putnam figures
among the latter, and considers the inseparability of the descriptive and
the normative aspects in thick ethical concepts as proof of the untenability
of the fact/value dichotomy, since the irreconcilable contrast that is often
emphasized between ethical statements and empirical descriptions would

turn out to be nothing more than a linguistic version of the dichotomy. In his view, an adjective like "'cruel' simply ignores the supposed fact/value dichotomy and cheerfully allows itself to be used sometimes for a normative purpose and sometimes as a descriptive term" (Putnam 2002a: 35), without the evaluative aspect and descriptive aspect being able to be clearly separated in all these uses. If, for example, I describe a teacher by stating, "'He is very cruel,' I have both criticized him as a teacher and criticized him as a man. I do not have to add, 'He is not a good teacher,' or, 'He is not a good man.' [. . .] Yet 'cruel' can also be used purely descriptively" (Putnam 2002a: 34). Even the historian's statement in the example above, the one that proclaims "He was a very cruel monarch," can both evaluate and describe,

> but I doubt that the *purpose* of historians would normally be to perform the speech-act of "condemning" long-dead persons; their aim is usually to make the historical events intelligible, and to do this they employ descriptions, which are themselves made available by a moral point of view—this is just the phenomenon of "entanglement" (Putnam 2003: 116).

What Putnam wants to call attention to is that

> there are facts [. . .] which only come into view through the lenses of an evaluative outlook. [. . .] If we define a "cruel" person merely as one who causes unnecessary pain (or causes pain out of malice), and take pain to be merely physical pain, then we shall miss all of the subtler forms of cruelty and all of the subtler motives for, and rationalizations of, cruelty (Putnam 2003: 112).

In other words, if we took only the descriptive side of "cruel," and said that this adjective means "the cause of deep suffering," we could not distinguish between facts in which cruelty is actually present and facts from which it is absent:

> Before the introduction of anesthesia at the end of the nineteenth century, any operation caused great pain, but the surgeons were not normally being *cruel*, and behavior that does not cause obvious pain at all may be extremely cruel (Putnam 2002a: 38).

Similarly, "willing to take serious risks" does not identify a person as "brave," since it is a definition that also applies to people who in everyday language are called "crazy," and facts about the latter would not be distinguishable from facts about people like the former. Therefore, since "every fact is value loaded and every one of our values loads some fact" (Putnam 1981: 201), we have that without values not only could we not have a physical world, as we saw in the previous section, but we could not even have a human world:

> The world we inhabit, particularly when we describe human beings for purposes other than the purposes of physics or molecular biology or some other exact science [. . .], is not describable in "value-neutral" terms. Not without throwing away the most significant *facts* along with the "value judgments" (Putnam 2003: 112).

THE RELIGIOUS DIMENSION

Certainly for Putnam "the purposes of physics or molecular biology or some other exact science" are not the only purposes we must have if we are to get a reliable picture of the world we live in. As is now clear since we have come this far, this is an aspect of his thinking that characterizes his personal naturalist perspective. The adjective "personal" is as appropriate as ever here.

Indeed, in Chapter 7 we saw how in his view the world is composed of various levels of form and, as we are highlighting in this chapter, these levels include the moral level. In addition, Putnam argues, there is a *religious* level. And this is probably something that most naturalists—including liberal ones—are unwilling to admit for the simple reason that "religion is also [and I would add, *mostly*] a personal matter or it is nothing" (2008a: 1). If this is indeed the basic reason why religion is considered incompatible with naturalism, the last quote shows that Putnam is not only aware of this but, as it were, takes the bull by the horns and immediately addresses the philosophically thornier issue, since: "I am still a religious person, and I am still a naturalistic philosopher" (2008a: 5). So, in what sense can religion concern not only the personal world but also the world in which we all live?

The answer lies precisely in the centrality of *practice* that we have seen in the previous sections to be at the heart of Putnam's ethical conception.[9] Interestingly, one of his main references in this regard is Pierre Hadot, a French philosopher and historian of ancient philosophy whom we have already met briefly. Hadot was the author of a book with a very telling title—*Philosophy as a Way of Life*—in which he invites us to look at ancient philosophers not as aiming at the production of theories and philosophy as an academic discipline, but rather as people committed to finding ever better ways of dealing with the problems of everyday life and (philosophy) as not only "a specific type of moral conduct" but rather

> a mode of existing-in-the-world, which had to be practiced at each instant, and the goal of which was to transform the whole of the individual's life (Hadot 1981: 265).

And, we note in passing, to illustrate such an idea of philosophy Hadot used not his own words but drew on a first-century *Jewish* philosopher, Philo of Alexandria—as if to say that if philosophy is to be conceived as a mode of existing-in-the-world, religion can be seen as intrinsically linked to this mode (cf. Putnam 2005b). Indeed, "religion is one participant in the ongoing process of moral inquiry" (Ruth Anna Putnam 2015: 722).

Regarding this conception of philosophy, Hadot is not Putnam's only point of reference. Alongside the ever-present Wittgenstein there also appear Søren Kierkegaard,[10] Martin Buber, Franz Rosenzweig, and Emmanuel Levinas. All of them "are thinkers who very much represent the ancient tradition that Hadot writes about" (Putnam 2008a: 13), and all of them help highlight a very important consequence of this conception of philosophy: a marked downsizing of the value of *theories*. But, it comes to be asked, what is wrong with theories? Isn't the knowledge we possess in a wide variety of fields *also* expressed through theories?

Generally speaking, a "theory" can be said to be something rigid, an explanation of a given set of phenomena that, while eminently hypothetical and fallible in character, implicitly raises the claim to constitute *the* correct and complete account of the set in question, whatever it may be. Such rigidity, however, clashes with the fluid and changing character of reality: from this point of view, as Wittgenstein repeatedly warned, a theory represents nothing more than an effort to harness a matter that continually escapes the theoretical grids one wishes to impose on it. In addition, a theory is itself something "abstract," although in some cases it may deal with something very concrete, and this abstract/concrete opposition is ineliminable.[11] Therefore, when philosophical thought deals with religion, this comes to the fore particularly in the impossibility of *theorizing* about God, about religious belief, about faith. This is where Buber, Rosenzweig, and Levinas come in: for them, the move away from theorizing coincides with a move toward common sense and a philosophical reflection in which religious sentiment is closely intertwined with everyday life, delineating itself as a way of life that delves deep. The value of religion—as well as the value of any genuine philosophy—resides then in fostering an intimate transformation of one's life: not the attempt to "reason" about God as an entity endowed with an essential yet inscrutable nature, but rather the development of the ability to cope as best we can with the myriad of existential problems that life inevitably confronts us with.

This being so, it is easier to understand how the religious dimension can be made compatible with a naturalistic perspective. If it is not a matter of theorizing about God—as this would implicitly lead to postulating both His actual existence and that of other supernatural entities—seeking by this route to find meaning in human existence, then the basic requirement of naturalism is safeguarded.[12] What Putnam is telling us is that to be a religious person one does not have to situate oneself on a metaphysical level and adopt an "onto-theological" position that leads one to believe in the existence of supernatural *entities*—in particular, one does not need to have a "metaphysical concept of an impersonal God, let alone a God who is 'totally other'" (Putnam 2008a: 102). On the contrary, it is

> possible to have a deeply fulfilling religious attitude while keeping far away from metaphysics. Speaking for myself I would say that although I do conceive of God as a "transcendent being," as a "necessary being," as an "unconditioned ground for the existence of everything that is contingent," I feel that insofar as I have any handle on these notions, I have a handle on them as *religious* notions, not as notions that are supported by an independent philosophical *theory* (Putnam 2005a: 579).[13]

Again, no "theory" is at issue here: if I am convinced of the existence of God, this does not happen because some (metaphysical) theory persuades me of His existence—maybe a theory that contains a "proof" of that existence—as is generally the case with the classical theist. A theory is of no use here since "the role of religious experience is not to *prove* something but to confront one with an existential choice, to make 'believe or don't believe' a 'live option,' in William James's words" (Putnam 2005a: 580–581). One can therefore be a religious person and not believe in anything like a "'first spiritual cause'" (Putnam 2015i: 706)—that is, in short, be an atheist. "I was a thoroughgoing atheist, and I was a believer" (Putnam 1992b: 1) wrote Putnam in the early 1990s speaking of his former self, but the verb would also be suitable in the present tense.[14] But, the question arises, is there not a contradiction in claiming, on the one hand, to be an atheist and, on the other hand, to believe in God as a transcendent and necessary being?

Not if one understands God in the correct way. And for Putnam this way lies "between John Dewey in *A Common Faith* and Martin Buber" (Putnam 2008a: 100). From the fact that in that book Dewey speaks of God as a human projection embodying our highest ideals, it follows that God is a human construction and that "the kind of reality God has is the reality of an ideal" (Putnam 2008a: 101). Now, it is true that both constructions and ideals are something subjective, in the sense that it is we human subjects who conceive them, but this does not imply that they are bound to be necessarily confined to a subjective sphere and thus to be "unreal." Just think of ideals such as those of freedom and justice, and how many real-world reactions and changes they have provoked throughout history:

> if these ideals had not at times been overwhelmingly "real" to some individuals, notwithstanding the circumstance that they are woefully far from being *realized*, this would be a far more intolerable world that it is (Putnam 2008a: 101–102).

This shows not only that, in spite of their subjective origin, ideals can be analyzed and discussed just as we find in any sphere in which one can argue rationally and arrive at conclusions endowed with objective validity, but also that this is possible because such objectivity can rest on something that is independent of us. This is why Putnam states that "which values and ideals enable us to grow and flourish is not a mere matter of 'subjective opinion'; it is something one can be right or wrong about" (Putnam 2008a: 101), and objectively so, because

> we construct our images of God in response to demands that we do not create, and [. . .] it is not up to us whether our responses are adequate or inadequate (Putnam 2008a: 6).

Dewey thus helps to understand that God is real just as certain ideals are real. However—and this is where Buber comes in—"God is not an ideal of the same kind as Equality or Justice" (Putnam 2008a: 102). It is something different, for God can be understood as a person, a "special" person in the sense of being supremely wise, kind, and just. And if God can be understood as a person—although, in light of the above, not "literally"—then it is possible to have a dialogical relationship with Him by entering into that relationship that Buber called "I-Thou." And that is all it takes to believe that God is a transcendent and necessary being while denying that He exists in a supernatural dimension necessarily beyond the reach of human knowledge. "Putnam, using terminology he learned from Stanley Cavell, suggests that believers do not proclaim their knowledge of God's existence but rather *acknowledge* Him" (Ruth Anna Putnam 2015: 708; my emphasis).

So, for Putnam, God is "neither an impersonal God arrived at by logic or metaphysics nor a wholly other God far removed from us" (West 2015: 766), yet He is present in a religious experience that is capable of conferring meaning on human existence. And this is because the meaning of our existence is not

> the meaning of "another life" but that of this our life, not that of a "beyond" but of this our world, and it wants to be demonstrated by us in this life and this world (Buber 1923: 159),

without the need to refer to some supernatural dimension. For Putnam, religious experience can be something quite common, such as

> the way in which a religious person may at any time experience something or some event, whether it be an obviously significant one, say, the birth of a child or the sort of deep crisis in one's life that William James describes in *The*

Varieties of Religious Experience, or a superficially "ordinary" one, as full of religious significance (Putnam 2005a: 568).

From this point of view, spirituality appears both *miraculous* and *natural*, as long as one does not regard miracles as the result of the intervention of a "supernatural helper": spirituality can be miraculous and natural

> just as the contact with another in what Buber calls the "I-You" relation is miraculous and natural, and the contact with natural beauty or with art can be miraculous and natural (Putnam 2008a: 102),[15]

and "both religious experience and aesthetic experience can be *enriched* by reflection" (Putnam 2015l: 726), where "spiritually enriching reflection differs from mere 'theorizing'" (Putnam 2015l: 728), the latter being as we now know beside the point in such a context.

> Theorizing about the ontological status of God, or the way of Heaven, or the Tao, or the Hindu Gods who are simultaneously six and thirty-three and 3,306 in number, is beside the point. What one wants to talk about is the ways in which one's life can be informed by asking what God wants of one, or asking how one can be in tune with the way of Heaven, or with the Tao (Putnam 2015l: 728).

Having a fulfilling and genuine spiritual dimension to one's life does not then necessarily involve thinking in terms of "God," and the writings of all three Jewish philosophers help Putnam substantiate his own religious position. All three are "impressive" (Putnam 2008a: ix), but in the end the one with whom the attunement is full is Buber himself, an author for whom Putnam makes no secret of his own predilection. Although in *Jewish Philosophy as a Guide to Life* the shortest chapter is devoted to him, he is the only philosopher toward whom no critical reservations are ever expressed, but whose thought is on the contrary defended against detractors and possible misunderstandings. In contrast, there are aspects in the other two thinkers with which Putnam strongly disagrees: in particular, in Rosenzweig the ontological thesis of a clear distinction between God, Man, and World, and in Levinas the thesis that God requires of us a total willingness to help the other, which makes the relationship with the other *asymmetrical*:

> the "asymmetry" of the ethical relation need not be carried as far as Levinas carries it. And—incorrigible Aristotelian that I am—I would not carry it that far. It is, I think, because Levinas thinks of ethics as the *whole* of "the true life" that he does so. But to be *only* ethical, even if one be ethical to the point of martyrdom, is to live a *one-sided* life (Putnam 2008a: 97–98).

Therefore, "'total responsibility' for the other seems to me to go beyond what is right to demand" (Putnam 2008a: 120). To live a fully ethical life, it is enough to enter into an "I-Thou" symmetrical Buberian relationship with God, which, when truly established, cannot but bring with it an enrichment of ethical relationships with fellow human beings: "if one is fortunate enough to experience an 'I-Thou' relation to God, then, Buber tells us, one will find all one's other (positive) I-Thou relations 'taken up' in it" (Putnam 2008a: 104).

In any case, what matters most to Putnam is that all three Jewish thinkers emphasize, each in his own way, that the function of religion is inner transformation in one's life, just as the ancient philosophers discussed by Hadot thought of philosophy: a "'philosophy as a way of life' [. . .] but only when it leaves the page and becomes 'experiential'" (Putnam 2008a: 108). Here then, finally, is how spirituality can remain within the boundaries of naturalism. Let Ruth Anna Putnam close the section:

> The temptation [. . .] to see a conflict between, as people say, science and religion vanishes when one realizes that theorizing about God is beside the point, that one can theorize only about the relation, or the interaction, between the human person and God, and, it will turn out, the relation between one human person and another (Ruth Anna Putnam 2015: 712).

RUTH ANNA: PARTNER IN LIFE AND PHILOSOPHY

Ruth Anna Putnam, Hilary's "brilliant and wise wife," was "a distinguished philosopher in her own right and beloved Professor of Philosophy at Wellesley College for decades" (West 2015: 766). They met on August 27, 1960, at a reception in Rudolph Carnap's honor at the International Congress on Logic, Methodology and Philosophy of Science in Stanford,[16] and they married on August 11, 1962. She was a scholar of pragmatism, and not only did she have an influence in her husband's approach to pragmatism, as we shall see in Chapter 9,[17] but developed a reflection on the entanglement of fact and value that was intertwined with her husband's in virtually inseparable ways, as will be evident from what follows.

Indeed, Ruth Anna Putnam relentlessly maintained throughout her career that "there is no epistemologically interesting difference between facts and values," given that "facts and values emerge at the end of inquiry on an equal footing: the facts are value-laden, and the values are fact-laden" (R. A. Putnam 1998a: 408). She argued that the conviction according to which there is an "interesting" difference between facts and values—*of course* there is a difference but, as we have seen from her husband's writings, it responds to no essential difference in the character of the respective judgments—stems from convictions such as (in ethics) that there are no absolute moral facts, and hence no objectivity in ethics, which is in turn motivated by the contrast between the certainty of scientific knowledge and the uncertainty of moral beliefs. Now, whilst noting that non-scientists tend to exaggerate the certainty of scientific knowledge, she makes a twofold comment: on the one hand she grants that there are no absolute moral facts but stresses that there are no absolute non-moral facts either, and on the other hand she highlights how there are grounds enough for arguing in favor of the objectivity of value judgments—and in ways not dissimilar from the way one justifies the objectivity of judgments of fact.

Before examining the two horns of her comment, it is worth mentioning her advice to take talk of facts with due caution. A quick glance at this talk brings to light that "the notion of a fact, or the associated notion of a factual statement, is hopelessly fuzzy, quite aside from any possible relation to values" (R. A. Putnam 2017: 105). We usually contrast facts with theories, but—she asks—where are we to draw the line between a theory and the facts it explains? This is a question with no answer, as our linguistic practice reveals. We often say something like "Evolution is a fact" to which our opponents may retort "Evolution is (just) a theory." Moreover, what is *actually* a fact may vary with the context. The facts explained by scientific theories are generalizations concerning repeatable events, not particular events; yet in courtrooms judges want the witness to report a particular event, that is, the fact that she or he observed. And, as far as factual statement are concerned, they usually have a hypothetical nature, which is made clear by statements such as "(I believe that) her friend is divorced"; yet we say also something like "In fact, he is divorced." All this shows that whether we classify an item as a fact, a theory, a conjecture, a generalization or a particular, depends on our *purposes* and *interests*—it is not given once and for all. No great philosophical conclusion follows from these examples, Ruth Anna Putnam admits: just the suggestion that "it might be wise to eschew talking about facts or factual statements as if these notions were clear" (R. A. Putnam 2017: 106).

Rather, an interesting philosophical conclusion may arise if we appreciate that absolute independent non-moral facts don't exist. Consider again the distinction between theories and facts, and let's exploit it to cast doubts on the notion of facts *independent* of us. We commonly draw this distinction by saying that theories explain facts, where the implicit assumption is that these facts are mind-independent. However, first-rate theories incorporate and express a good deal of creativity, where this means that they are not mere summaries of old facts but new ways of looking at old facts, bringing to light both similarities and differences among the things being analyzed that no one had been able to see before. An analogy may clarify the point: the analogy between theory-construction and a child asked to sort pieces of cardboard.

Some ways of sorting will be accepted by us immediately; they fit our notion of "sorts" (color, shape, size, texture), and others we may accept if the child can give a reason (red and yellow pieces go together because these are warm colors, blue pieces are separate because they are cold). Some sortings will be so bizarre from our point of view and so inexplicable by the child that we are going to regard them as "wrong" (R. A. Putnam 1985: 392).

This may surely be taken to exemplify how theories are creatively conceived, thanks to the child's new ways of looking at old facts, but "it is perhaps better to consider it as an example of fact-making" (R. A. Putnam 1985: 392), in that the child didn't explain old facts—as theories are supposed to do—but "created" facts of similarity. Now, despite the potentially misleading character of fact-making talk, Putnam's point is clear: "creativity consists in 'seeing and responding' as opposed to cognizing and verbalizing. [And this kind of seeing] consists in becoming aware of what has been excluded from notice" (R. A. Putnam 1985: 397)—features of the world that, strictly speaking, come into existence only when somebody gives enough reason to make us contemplate them. Therefore, "creation" has a metaphorical flavor here, since we are no gods who make the world appear just like that, but it helps to become aware that a good number of facts are mind-dependent, relative to us, theory-laden, precisely because theory-making and fact-making turn out to be intimately interwoven. But this does not jeopardize the *objectivity* of facts. Why not? Because this kind of fact-making is not arbitrary, and what keeps it from being arbitrary are elements that are as independent of us as the objects of sensory stimulation, past tradition, and the implicit and explicit criteria for consistent theory-construction. Indeed,

(1) we are constrained by the actual sensory inputs we receive; (2) we are constrained by what we have made of our sensory inputs in the past, the conceptual framework embodied in our prior beliefs, what we have taken to be facts so far; and (3) we are constrained by the insistent demand for coherence and consistency (R. A. Putnam 1985: 397),

all constraints that place a limit on our creativity and make its results non-arbitrary. Above all, the arbitrariness of fact-making is averted by its responding to *needs* and *interests* that are not exclusively of our making but come from an independent world and, again, constrain our theorizing. This may become clear if we reflect on the fact that our focusing on this or that aspect of reality may be driven by a newly arisen need to solve a given problem and our interest in solving it, where needs and interests may be both theoretical and existential in nature. But this need isn't produced by us on a whim, and the correctness of the solution to the problem we might put forward isn't determined by us. As Ruth Anna Putnam emphasizes, "our interests legitimately determine the subject matter of our beliefs—what we investigate—but not the result of those investigations" (R. A. Putnam 2017: 110). And the latter results, although they bear an inevitable human imprint, are so far from being arbitrary as to count as *facts*. Facts then (even non-moral facts) aren't absolute and completely mind-independent.

So much for the first horn of Ruth Anna Putnam's comment on one of the central statements of the supporter of the fact-value dichotomy—the statement that, owing to the non-existence of absolute moral facts, there is no objectivity in ethics. Regarding the second horn, her thesis is that we seek and obtain objectivity along paths that are the same in both the natural and the human sciences. Her argument exploits what we have just seen, and so my illustration of it can be brief.

That we create values of any kind (moral, aesthetical, political, etc.) is a common view, but again this doesn't necessarily make them arbitrary. What has been observed about creativity in the natural sciences applies here too. As we have seen also maintained by her husband, values are created in response to needs that we don't create, first of all the need to have values themselves: "we need values in order to have reasons (of the kind that may serve as justifications) for our actions, and more generally in order to choose [. . .] that is, when neither instinct, nor deeply ingrained habit, nor internalized values, nor a set of explicit instructions has set us on a road with no exits" (R. A. Putnam 1985: 399). And specific constraints arise with these needs:

> when we make moral values because we need to live in a human society, because we need to cooperate with each other at least somewhat, this need generates some constraints: some lives must be "sacred," some agreements must be trustworthy, some originality must be permitted, etc. (R. A. Putnam 1987: 74)

From this there arises a detailed moral-legal-political structure, whose open-ended texture is constantly expanded and whose various parts are constantly revised as the need arises, providing as firm an anchor for our behavior as we humans could ever aspire to. And this—far from being the objectivity of a God's Eye View—is objectivity enough:

> Just as fact-making and theory-making turn out to be intimately interwoven, but facts are nevertheless *solid* enough to allow us to navigate a perilous world, so basic values and detailed moral structures are intimately interwoven and *solid* enough to enable us to navigate the perils of human relationships (R. A. Putnam 1985: 402–403, my italics).

OBJECTIVITY

The main consequence that the interweaving of facts and values has on the epistemological level is the opening of *rational argumentation* in every direction: argumentation aimed at showing the correctness of the theses, claims and theories we support—their possible truth. If facts and values are inextricably intertwined, then everything that has traditionally been considered valid only in the realm of facts also applies to the realm of values. In particular, notions such as truth, reference, justification, and correctness find application not only in the natural sciences, but also in the realm of the so-called human sciences. Even in the latter—and thus in ethics, aesthetics, economics, politics, sociology, philosophy, law, and so on—it is possible to arrive at genuinely cognitive results endowed with *objective* validity and susceptible of being considered true or false. As for ethics, we have that "for reasons internal to ethics itself, ethical truth cannot be 'recognition transcendent'" (Putnam 2015f: 797): that is, moral truths, like aesthetic truths, seem to be constitutively "human-friendly."[18]

Thus, the language of the humanities is also the language of objectivity. In fact, "it is time we stopped equating *objectivity* with *description*," stopped believing "that 'objectivity' means *correspondence to objects*" (Putnam 2002a: 33), since in several areas of human knowledge we have truths that do not presuppose any reference to objects: this is the case with mathematics, logic and, in light of the discussion above, ethics.[19] As for the latter, the objectivity of ethical principles

> is connected with such things as width of appeal, ability to withstand certain kinds of rational criticism [. . .], feasibility, ideality, and, of course, with how it actually *feels* to live by them or attempt to live by them (Putnam 1976b: 93).

There is no doubt that ethical principles and values are a human creation, in the sense that it is we who establish the values—*religious values* included—to which we wish to subordinate our conduct, and who formulate the norms intended to regulate life within a given collectivity of human beings by finding the most suitable solutions to the problems generated by coexistence; and there is no doubt that such principles and values change and evolve throughout history, as our species acquires additional discernment relating to human and non-human reality, thereby refining its capacity for orientation. But subjecting principles and values to rational criticism shows in itself that there are better and worse ways of responding to problematic situations (cf. Putnam 1981: 163), better and worse ways of formulating principles and values, and thus shows the existence of facts of the matter—not determined by us but by reality—that make certain value judgments true and others false (cf. Putnam 2002a: 108). If, therefore, on the one hand "we *make* ways of dealing with problematic situations," on the other hand "we *discover* which ones are better and which worse" (Putnam 2002a: 97, italics mine). Reality is thus not morally indifferent: it solicits responses from us, in terms of values and norms, and then "determines whether our responses are adequate or inadequate" (Putnam 2008a: 6).

Far from having a justification external to the conceptual systems that vivify cultures coexisting in space and succeeding one another over time, a justification imposed by a fixed and prefixed reality, objectivity is for Putnam "an ongoing *achievement*—one that must be constantly rethought" (Bernstein 2005: 260): an objectivity with a human face that concerns facts as much as values.[20] And if all the objectivity to which we can aspire can only be "objectivity humanly speaking, it is still objectivity enough" (Putnam 1981: 168), for it is "the objectivity of what is objective from the point of view of our best and most reflective practice" (Putnam 1994b: 177).[21]

This objectivity is therefore the concern of philosophy—and in all fields of our intellectual enterprise, as this book tries to show. Indeed,

> philosophy does not spring up in a void. Great philosophical movements arise from reflection on life and on the place of humanity in the world. Again and again they have proposed ways of redirecting both individual and social life. This activity—the activity of putting forward and discussing what I called "moral images of the world" in *The Many Faces of Realism*—seems to me *the* indispensable task of philosophy. Philosophy certainly needs moments of technical argument, and it needs moments of exposing nonsense, but neither of these adds up to anything of lasting value in the absence of moral imagination (Putnam, 1992i: 376–377).

NOTES

1. Both passages find their combination a few years later in the following statement made in connection with the thought of Pierre Hadot, a historian of ancient philosophy with whom Putnam shows great sympathy: "the ancient idea of transforming one's way of life and one's understanding of one's place in the larger scheme of things and in the human community is one that we must not lose. Philosophy certainly needs analysis of arguments and logical techniques, but is in danger of forgetting that these were originally in the service of this very idea" (Putnam 2008a: 13). This is a conception of philosophy that Putnam never abandoned, and it is closely related to his liberal naturalism, as is clear from the following passage: "I have remarked in more than one place that philosophy needs both a technical side, a side that requires rigorous arguments, and a less technical side, a side that seeks to inspire us to live more fulfilling lives, but not necessarily by offering the sorts of arguments that can be formalized (although the choice of examples by a great moral philosopher can itself

represent a form of argument). Without the technical side, philosophy risks becoming homiletics; without the inspirational side, it risks falling prey to the illusion that philosophy can and should become a science. (Moreover, the idea that the inspirational side should disappear today reflects a noncognitivism with respect to values that seems to me bad philosophy, an intellectual mistake.)" (Putnam 2015h: 678). Criticism of noncognitivism with respect to values will be addressed in this chapter.

2. This position can be traced back to the meta-ethical particularistic conception: see Dancy (2017).

3. As Putnam's wife Ruth Anna, a renowned moral philosopher, pointed out, "the notion of value is enormously rich. Various kinds of things (objects, states of affairs, character traits, etc.) have value, are valued, are evaluated, etc. Value may be monetary, moral, social, epistemic, aesthetic, etc. None of these lists are exhaustive" (R. A. Putnam 2017: 106).

4. In maturing Putnam's idea of the untenability of the dichotomy between facts and values, a central role was played by the reception of the lessons of the classical pragmatists, as we shall see in the next chapter. Another important role was played by the exchange with Jürgen Habermas and Richard Rorty (on the latter exchange, cf. Chakraborty 2019).

5. On the fusion of facts and values see also Elgin (1996, 2007).

6. In this regard, Ruth Anna Putnam argued that "the crucial role epistemic values play in science [. . .] was, in my opinion, Hilary Putnam's most important early contribution to the philosophical debate concerning the status of values" (R. A. Putnam 2013: 240).

7. The role of the aesthetic value of beauty in scientific inquiry was emphasized by English physicist Paul Dirac, who thought that "a theory with mathematical beauty is more likely to be correct than an ugly one that fits some experimental data" (quoted in Hovis & Kragh 1993: 104). See also Putnam 2002a: 31.

8. From a terminological point of view, thick and thin concepts began to be discussed with Ryle (1968) and Geertz (1973a); driving the debate in metaethics, however, was Williams (1985). For a general overview see Kirchin (2013).

9. Cf.: "Putnam's view of religion is largely a view of the ethical life" (Miguens 2020: 404).

10. From Kierkegaard, Putnam borrows a deeply reflective and deeply doubting religious attitude: "I [. . .] think that Kierkegaard was right to describe faith as something that is in a dialectical relation with doubt" (Putnam 2015l: 729). Cf. also: "Avi Sagi once told me that in a still-unpublished fragment—I think it was a diary—of Kierkegaard's, he found the words 'Leap of faith—yes, but only after reflection'" (Putnam 2005a: 567).

11. For an interesting critique of "theory" in the field of literary criticism, very much akin to Putnam's critique, cf. Compagnon (1998).

12. Reflecting on Putnam's religious position, John Haldane rhetorically asked, "if naturalism is sufficiently expansive to accord objective standing to the contents of moral and aesthetic experience, on what non-question-begging basis can it refuse recognition to the contents of experiences of spiritual transformation?" (Haldane 2015: 700)

13. In an attempt to reconcile naturalism with a religious position of a "metaphysical" character (which he judged to be indispensable for such a position) John Haldane proposed a distinction between the term "supernatural" and "what the medievals would have described as the 'preternatural,' that is, as being outside or contrary to the course of nature, as in someone or something levitating or remaining in a fire while showing no effect of heating. The supernatural, by contrast, belongs to the spiritual order and involves grace, divinely aided personal transformation, and other spiritual effects," and added, "While the preternatural may sit ill with Putnam's naturalism it is not at all clear that the supernatural must also do so" (Haldane 2015: 699). But, *pace* Haldane, it is on the contrary very clear: the "divine aid" that Haldane sees in the supernatural would be possible through an entity that would contradict the epistemology and metaphysics underlying the natural sciences.

14. Cf.: "in Rosenzweig's eyes I would have counted as an 'atheist theologian'" (Putnam 2008a: 103).

15. The analogy between religious experience and aesthetic experience is quite telling, and Putnam returns to it several times. Both types of experience are made possible by what Immanuel Kant called "indeterminate concepts": "Indeterminate concepts are not purely intellectual concepts; they require both a sensible subject matter and an active imagination to apply. That perception is fused with conceptual content is something we learned from *The Critique of Pure Reason*; that some of the perceptions we value most are fused with indeterminate, open-ended, conceptual content, content in which imagination and understanding cooperate under the leadership of imagination, is something we learn from *The Critique of the Power of Judgment*" (Putnam 2005a: 575–576).

16. "We still celebrate August 27 every year as the anniversary of the day we met" (Putnam 2015b: 56).

17. Ruth Anna "more than anyone else brought me to an appreciation of American pragmatism in general, and of the significance of John Dewey's contribution to philosophy in particular" (Putnam 2004a: ix).

18. I thank Stefano Caputo for suggesting this expression. For a discussion of Putnam's arguments for moral objectivity, cf. Timmons (1991).

19. This is obviously the sense of the title of Putnam's book *Ethics without Ontology*. On this aspect, cf. Laugier (2020).

20. Putnam's interpretation of objectivity thus stands as an alternative to two equally insidious opposites: those that a great thinker much admired by Putnam believed "the Scylla of timeless, abstract principles that [take] no account of life and change, and the Charybdis of relativism that destroy[s] morality or reduce[s] its goals to matters, in the end, of subjective temperament or inclination" (Berlin 2000: 257).

21. The breadth and significance of the notion of objectivity in Putnam are brilliantly discussed in Macarthur (2018).

Chapter Nine

Learning from Pragmatism

A SLOW APPROACH

In Chapter 8 we saw Putnam's conception of philosophy and how important to him was the moral aspect of our existences—and, consequently, how wide should be the space that philosophical reflection should reserve for it. The highly personal character of Putnam's philosophizing is once again striking: a thinker who not only made significant contributions to various areas of what is still called "analytic philosophy"—although it is an area of contemporary research that is evolving so inexorably that its original meaning has faded—but who also broadened his interests by turning to different traditions of contemporary thought. I see this as a clear indication that a great philosopher does not necessarily belong to a movement, but deals with whatever topics and philosophers interest them regardless of their membership in some defined movement.[1] Indeed, a movement, whatever it may be, cannot but constitute a conceptual grid with well-marked—and therefore in some cases too stringent—limits. And of this Putnam was supremely aware:

> movements are a bad idea in philosophy. When you think of philosophy as movement A against movement B you overlook both the insights of the movement you oppose and the errors of the movement you favor. [. . .] The thing about philosophy is that movements lead to slogans and then to frozen positions, which is of no benefit to anyone. [Therefore,] I do not see myself as a member of any tradition (Putnam 1997: 156).[2]

In particular, he did not consider himself a pragmatist—"I am more of a realist than a pragmatist" (Putnam 2015m: 768), or "I am a pragmatist in the sense that I am a liberal naturalist" (Putnam 2015o: 6)—since being a pragmatist in his view is tantamount to upholding a well-defined theory of truth—in Peirce's version, or James's or Dewey's.[3] But the very attempt to define the concept of truth is to be considered a philosophical error to be kept at bay: at most, as we saw in Chapter 7 (145ff), one can make "attempts to say something about its logical properties and connections with other concepts" (Putnam 1997: 155), and the form of alethic pluralism he advocates is intended to represent just that. Yet not only is much of what we saw in Chapter 8 due to pragmatism, but Putnam himself is often considered "a leading neo-pragmatist" (Macarthur 2017: 2). So, how do things stand?

Things stand very simply: although "I am not a pragmatist, [. . .] I have learned much from pragmatism" (Putnam 1997: 154).[4] The theses set forth in much of Chapter 8 are precisely a result of that learning. It is interesting to note that the Putnamian path to pragmatism was very gradual, and that it even began with a kind of hostility toward pragmatists—"I became hostile to pragmatism during my senior years at 'Penn'" (Putnam 1997: 153–154)— dictated by his youthful approach to logical positivism. Nevertheless, Putnam took a course at the University of Pennsylvania taught by an "atypical pragmatist," C. West Churchman, who reported the thoughts of a student of William James who later became emeritus professor at Penn, E. A. Singer Jr., and wrote the following propositions on the blackboard: *Knowledge of facts presupposes knowledge of theories* ("here Singer was attacking the idea that science can 'start' with bare particular data and build up to generalizations by induction and abduction. We always already presuppose a stock of generalizations when we observe," Putnam 2017: 19); *Knowledge of theories presupposes knowledge of facts* ("there are no generalizations about the world we can know a priori," Putnam 2017: 19); *Knowledge of facts presupposes knowledge of values* ("this is the position I defend," Putnam 2017: 19, and it "is as controversial today as it was then," Putnam 1992a: 14); *Knowledge of values presupposes knowledge of facts* ("Against all philosophers who believe that (some part of) ethics is a priori," Putnam 2017: 19; cf. also 1997: 150; 2015b: 11; R. A. Putnam 2013: 246ff).

Although West Churchman's course failed to enlist Putnam to the cause of pragmatism,[5] something must have remained if decades later he is able to recall the four propositions so vividly. The fact of the matter is that not even a seminar on Dewey's *Logic: The Theory of Inquiry* that he attended years later when he was a graduate student at University of California at Los Angeles succeeded in bringing about a serious interest in pragmatism, although Donald Piatt—the professor who taught the course—is later remembered as an excellent expert in Dewey and a good teacher. It was around the 1980s that Putnam began to take an interest in James and Dewey: in the former thanks to Richard Rorty and Jacques Rudner's book *A Stroll with William James* (Rudner 1983), which Rorty had reviewed favorably; in the latter—as we know—thanks to his wife, for decades an assiduous scholar of classical pragmatists and of Dewey in particular. From then on, the study of the writings of the pragmatists and courses devoted to them were a constant in Putnam's intellectual life, especially James and Dewey:

> In general, I found James useful for the big picture—the idea of the interpenetration of fact and value, for example—but when it comes to the details of how to proceed in ethics, eventually I found Dewey more useful (Putnam 1997: 154).[6]

Indeed,

> I subsequently returned several times to interpreting Dewey's ideas (particularly on the connections between ethics, democratic theory, and what he called the "logic of inquiry"). The heart of my interpretation can be found in my 1994 "Pragmatism and Moral Objectivity" (Putnam 2015b: 93).[7]

It is in this last essay that he declares:

> while I do wish to undermine moral skepticism, I have no intention of defending either authoritarianism or moral apriorism. It is precisely for this reason that in recent years I have found myself turning to the writings of the American pragmatists (Putnam 1994b: 152).

THE RELEVANCE OF PRAGMATISM

In Chapters 6 and 7 we followed Putnam's peregrinations into metaphysical territory, and we were able to appreciate the caution with which he describes the metaphysical position that characterizes his later period. We now discover that much of that caution stems from his rapprochement with pragmatism. Indeed, in "Pragmatism and Moral Objectivity" he makes clear that "turning to the writings of the American pragmatists" does not mean adopting a metaphysical theory plus a consequent theory of truth, but it does mean turning to a *group of theses* that formed the basis of the philosophies of the classical pragmatists, although each pragmatist examined and discussed them in his or her own way. It means *not* a rejection of metaphysics, but the delineation of a metaphysical conception that reflects our best and most reflective practice. Here, meanwhile, are the theses:

> (1) *antiskepticism:* pragmatists hold that *doubt* requires justification just as much as belief (recall Peirce's famous distinction between "real" and "philosophical" doubt); (2) *fallibilism:* pragmatists hold that there is never a metaphysical guarantee to be had that such-and-such a belief will never need revision (that one can be both fallibilistic *and* antiskeptical is perhaps *the* unique insight of American pragmatism); (3) the thesis that there is no *fundamental* dichotomy

between "facts" and "values"; and (4) the thesis that, in a certain sense, practice
is primary in philosophy (Putnam 1994b: 152).

All four theses underlie what we have seen in the previous chapters, either
explicitly or implicitly. The first is clearly expressed by the Putnamian argument against the BIVs hypothesis, the second animates the original Putnamian conception of necessity that we encountered in Chapter 1, while Chapter
8 contains the arguments with which Putnam argues for the last two. But it
is especially the fourth—the thesis of the *primacy of practice*—that plays an
ever more central role in Putnam's thinking, guiding his choice of themes
and philosophers to deal with. All of this starting roughly in the mid-1980s. I
share what James Conant argues in this regard:

> From the mid-1980s on, it is not an exaggeration to say that a whole new Hilary
> Putnam bursts onto the philosophical scene [. . .] His seminars, conversations,
> essays and books begin regularly to feature names which at most only rarely
> occur in his writings prior to the mid-1980s (Conant 2022: 36 and 38),

where among these names there are those of the classical pragmatists.[8] And
indeed a new argument that Putnam advances in the 1994 paper to support
the fourth thesis is linked to their names. At the end of Chapter 2 we briefly
encountered the so-called *indispensability argument* in favor of the existence
of mathematical entities such as numbers and sets, an argument associated
with the names of Quine and Putnam. Putnam now points out that, on closer
inspection, this argument is but one example of an appeal to practice made
within the philosophy of science, since—rather than arguing in favor of the
existence of numbers and sets from a metaphysical point of view—"Quine
deliberately displaces the discussion to a more humanly accessible region"
(Putnam 1994b: 153). Contrasting, on the one hand, a Platonist perspective
that the human mind has the possibility of "grasping" abstract objects existing in a dimension that is neither physical nor mental, and, on the other hand,
the thesis that we are able to interact causally with abstract entities, Quine
emphasized the indispensability of the conceptual scheme of set theory for
both mathematics and physical science, highlighting how

> what is indispensable to our best paradigms of knowledge cannot [. . .] be criti-
> cized from some supposedly "higher" philosophical viewpoint; for there is no
> "first philosophy" above and outside of science (Putnam 1994b: 153).

Now, Putnam argues, the same kind of argument can be traced in the writings of classical pragmatists—again, each in their own personal way—dedicated to the question of values:

Just as Quine displaces the traditional worry as to the mystery of our "interaction" (if any) with abstract entities such as sets and numbers, the classical pragmatists displaced the equally ancient worry as to the nature of our "interaction" (if any) with ethical properties and with value properties and normative properties generally (Putnam 1994b: 154).

Classical pragmatists, like Quine, also drew attention to the "function" performed by a certain kind of discourse—in their case discourse about values—emphasizing that this kind of discourse is "indispensable in science and in social and personal life as well" (Putnam 1994b: 154): in fact, value judgments are pervasive in all aspects of our lives, and, as we saw in Chapter 8, our practice shows that we do not regard them as a matter of mere taste; on the contrary, we argue in favor of them, defend them if we are criticized, try to show their correctness, using the *language of objectivity* in all this. Within value discourse, as well as within scientific, mathematical, and historical discourse, we use the same laws of logic and argumentative strategies. "And, like Quine, the classical pragmatists do not believe that there is a 'first philosophy' higher than the practice that we take most seriously when the chips are down" (Putnam 1994b: 154).[9]

This, then, is how the fourth thesis, that of the primacy of practice, is explained. A careful consideration of it reveals that an explicit or implicit "belief in the cognitive status of value judgments underlies an enormous amount of our practice (indirectly, perhaps, the whole of our practice)" (Putnam 1994b: 160). Moreover, a proper appreciation of this primacy and of *the agent point of view* that is what practice consists in cannot fail to have beneficial effects on our metaphysical conception. Far from favoring the traditional *third-person point of view*—aimed at obtaining an "objective" description of the world in the sense of impersonality, a description entirely devoid of human traces and hopefully complete—

> the pragmatists urged that the agent point of view, the first-person normative point of view, and the concepts indispensable to that point of view should be taken just as seriously as the concepts indispensable to the third-person descriptive point of view (Putnam 1994b: 168).

Of course, however, this exhortation can by no means be accepted by old-fashioned naturalists. It is well known that one of the most immediate reactions against the indispensability argument put forward by Quine was to argue that, even if we admit the indispensability of discourse about sets and thus their existence, we could not possibly *know* this existence or *refer* to sets, because this knowledge would require something else, such as causal interaction.[10] And thus, according to this objection, the indispensability argument would keep us on square one, with no real philosophical progress. The same objection is raised by those who, while admitting the indispensability of value discourse, believe that a realist perspective cannot in any case be adopted toward it,[11] and that therefore the argument of classical pragmatists would be unproductive. Putnam's response to this dual objection shows the pay-off of taking both pragmatism and practice seriously. Indeed, it is all too easy to discern a well-defined metaphysical option behind the objection: this, of course, in itself is not a problem, but it becomes a problem to the extent that this option is not adequately argued for and is simply *assumed*. Which makes it nothing more than a "metaphysical dogma." And that is precisely what happens here, since the objection seems to be motivated by the belief that only that which fits the image of the world provided by modern science is acceptable—scientific conviction far from being self-evident. What, therefore, appeared as a rational argument against the objectivity of discourse involving sets and against cognitivism in ethics reveals itself to be smuggling in a certain amount of suspect metaphysical baggage. This does not mean that pragmatism is "anti-metaphysical," but simply that it "goes with the criticism of a certain style in metaphysics" (Putnam 1994b: 159):

> Pragmatists do not urge us to ignore sound arguments against what we believe, when such arguments are advanced; they do urge us not to confuse the "intuitions" of metaphysicians with genuine arguments (Putnam 1994b: 156).

Now, it is very interesting to note that the letter of this last statement—the "letter," not the "substance"—would have been underwritten by Richard Rorty, another great philosopher who is usually referred to as a *neo-pragmatist*. I say "not the substance" because Rorty would turn such a statement in a decidedly anti-metaphysical direction. And several statements could appear indifferently in the writings of one or the other, despite their receiving different and sometimes opposing interpretations—"thus illustrating the well known maxim that one philosopher's *modus ponens* is another philosopher's *modus tollens*" (Putnam 1993: 280). Here is one example: Putnam's statement that

> elements of what we call "language" or "mind" *penetrate so deeply into what we call "reality" that the very project of representing ourselves as being "mappers" of something "language-independent" is fatally compromised from the very start* (Putnam 1990a: 28)

is enthusiastically endorsed by Rorty several times (cf. Rorty 1993: 43; 1994: 67–68; 1999: xxvii), and coupled with it is the following statement by Rorty that Putnam would have subscribed to without difficulty:

we will always be captive by some picture or other, for this is merely to say we shall never escape from language or from metaphor—never see either God or the Intrinsic Nature of Reality face to face (Rorty 1994: 80).[12]

Same wording, same words, same tone, but opposite philosophical interpretations. It is worth trying to understand why, especially considering that the classical pragmatists were an important source of inspiration for Rorty as well.

THE EXCHANGE WITH RORTY

Rorty's Epistemology in Outline

Richard Rorty can certainly be counted among the great philosophers of our age. A profound connoisseur of the history of Western philosophy, beginning at least with his masterpiece *Philosophy and the Mirror of Nature* (Rorty 1979) he began to subject the main trend traceable in that history to a close critique on the basis of a formidable battery of arguments.

Very succinctly, the main trend along which the history of Western philosophy evolved posits epistemology as one of the major tasks of philosophical inquiry and *mirroring* as the kind of epistemic relationship the human mind stands in to the world. From this perspective, mind and language aim to concoct *representations* of the world, and knowledge is gained when they manage to elaborate representations that are *true*. According to Rorty, this is a point of view that is *realist* in character and takes truth to be a sort of correspondence between a given thought or statement, and the piece of the world it is about. However, he found this perspective smacks of the untenable God's Eye View that Putnam variously criticized, as the world to which a purported representation is supposed to correspond is a *non-human* world, and as such inaccessible. Thus, the main trend of Western philosophy was bankrupt from the beginning, and the very idea of "representing" something is devoid of content. Consequently, he urged contemporary philosophers to stop worrying about the best interpretation of notions such as reality, truth, and objectivity, and to give up being "metaphysically active" (Rorty 1995: 29, 38 ff). In his views, it is philosophically much more fruitful to become more sensitive to the needs and demands human beings are up against on a daily basis, just as the classical pragmatists taught—and we cannot help noticing the consonance with Putnam on the latter point. This would be the only way to improve our situation in the world, achieving the highest possible degree of *solidarity* with our cultural peers, and setting aside the impossible quest for absolute objectivity.

Not that the world does not exist, since there is nothing to represent, or that there is no truth,[13] according to Rorty, or even that there can be no talk of objectivity: only, these notions must be wrested from the monopoly of philosophy and returned to their ordinary use. And so the "objectivity" to which the realist philosopher illusorily aspires must be replaced with "solidarity"—a notion certainly manipulable by us human beings—and truth with something equally manipulable such as warranted assertibility or, even better, with the simple use of the word "true." For Rorty, this means being pragmatists who correctly draw the lesson of the classical pragmatists. In fact, to be pragmatists means

> view[ing] truth as, in William James' phrase, what is good for *us* to believe. So they do not need an account of a relation between beliefs and objects called "correspondence," nor an account of human cognitive abilities which ensures that our species is capable of entering into that relation. [. . .] For pragmatists, the desire for objectivity is not the desire to escape the limitations of one's community, but simply the desire for as much intersubjective agreement as possible, the desire to extend the reference of 'us' as far as we can (Rorty 1985: 22–23).

An interpretation of truth *à la* James thus helps to see how objectivity can only be solidarity. It must be made clear, however, that, after oscillating between Jamesian and deflationary interpretations of truth, Rorty ended up embracing the latter. In the following passage Rorty explains why:

> Philosophers who, like myself, find this Jamesian suggestion persuasive, swing back and forth between trying to reduce truth to justification and propounding some form of minimalism about truth. In reductionist moods we have offered such definitions of truth as "warranted assertibility," "ideal assertibility," and "assertibility at the end of inquiry." But such definitions always fall victim, sooner or later, to what Putnam has called the "naturalistic fallacy" argument—the argument that a given belief might meet any such conditions but still not be true. Faced with this argument, we pragmatists have often fallen back on minimalism and have suggested that Tarski's breezy disquotationalism may exhaust the topic of truth (Rorty 1995: 21).

The naturalistic fallacy cited by Rorty is called the "idealistic fallacy" by Putnam, and we encountered it in Chapter 5. It is indeed a good argument against any attempt to reduce truth to some notion judged to be more plausible. Except that, while Putnam draws from such an argument a reason to regard truth as an irreducible and substantial notion, Rorty incorporates this argument into his own anti-metaphysical stance and treats it as a demonstration of the non-substantial character of truth, embracing a deflationary position—position that as we know revolves around the "equivalence thesis"—the version of which applied to sentences is due to Alfred Tarski. More specifically, Rorty's deflationary account of truth revolved around three theses concerning the use of the word "true." In his view, the uses to which this word can be subjected are no more than the following: the *disquotational*, the *commending*, and the *cautionary*. The first is the use based on the equivalence thesis—the one that equates an attribution of truth to a (quoted) sentence with the sentence itself (without quotes and the phrase "is true"). The second is the use of the word we make when we want to endorse a statement or pay an implicit compliment to its author (e.g., when we say "The statement she just made is true").[14] The last use of the word "true" is the one we make when we say that a statement is correct, but we acknowledge at the same time the possibility that it might be mistaken. In a case like this we might say something such as "This statement is fully justified, but perhaps not true"[15]

Here, then, is how Rorty is willing to talk about objectivity and truth. Both are simply what human beings determine they are. They are never fixed by something non-human, or by something independent of anybody's will. That is why "we need to restate our intellectual ambitions in terms of our relations to other human beings, rather than in terms of our relation to non-human reality" (Rorty 2000: 25). Justification of our intellectual ambitions cannot have anything to do with the world, because we humans can only interact with what is human or connected to it. Therefore, justification can only be obtained from other human beings—from *audiences*, as he puts it. For any claim (whatever the content) the criterion of rightness is the collective body of fellow speakers, not "impersonal" entities such as *the world*, for the simple reason that the world does not talk, and "if it does not talk, we are not answerable to it" (Rorty 2015: 864).

As for the world, Rorty had no problem recognizing that there is an independent world—for instance, quarks existing independently out there. And he had no problem recognizing that we human beings are "connected" to the world. Only, having ruled out as we have seen a representational relation, he believed that the only relation we can have is a *causal* relation. According to him, this is a relation describable in such simple but theoretical terms as those we get from Darwin's theory of evolution. The story is familiar. The natural environment plays a multifarious causal role that impinges on us in various ways, and at the same time we act within that environment causing a series of changes in it. In brief, the causal transactions we have with the world are the same as those that *animals* have with the world. And—according to Rorty—not only animals, but also artifacts such as *computers*. Here is what he says:

> human beings' only "confrontation" with the world is the sort that computers also have. Computers are programmed to respond to certain causal transactions with input devices by entering certain program states. We humans program

ourselves to respond to causal transactions between the higher brain centers and the sense organs with dispositions to make assertions. There is no episte-mologically interesting difference between a machine's program state and our dispositions, and both may equally well be called "beliefs" or "judgments" (Rorty 1998b: 141).

So, the idea is that if we want to know something about the causal link that characterizes our relation to the world—the link that would allow us to say the world is independent—we should go and see what Darwinians have to say about it (as well as neo-Darwinians, insofar as cultural evolution is concerned). But what we unavoidably get is nothing but a theory or descrip-tion or picture *we* advance, given that "we shall never escape from language or from metaphor" and never find a causal relation, so to speak, out there, to be looked at face to face. On the contrary, "the required notion of 'causal connection' is deeply *intentional*," as Putnam puts it, thus marking another point of convergence with Rorty (cf. for instance Putnam 1990c: 173), that is, something which reflects our interests, aims, and desires. Here is one of the passages in which Rorty makes this point:

[the] causal independence of giraffes from humans does not mean that giraffes are what they are apart from human needs and interests (Rorty 1999: xxvi).

Notice that Rorty is willing to pursue this point to its extreme conse-quences, consequences that threaten the very idea of independence. Here is a revealing quotation:

The causal independence of quarks from human discourse is not a mark of reality as opposed to appearance; it is simply *an unquestioned part of our talk about quarks.* [. . .] We can say, with Foucault, that both human rights and homosexuality are recent social constructions, but only if we say, with Bruno Latour, that quarks are too. There is no point to saying that the former are "just" social constructions, for all the reasons that could be used to back up this claim are reasons that would apply to quarks as well (Rorty 1998a: 8; my italics).

In other words, the distinction between what is socially constructed and what is "out there" is irrelevant. And the point is that both are part of the way in which we talk, and the way things are currently going is that in our talk we sometimes put into question the former (e.g., human rights) and not the latter (e.g., quarks). That is why we might (falsely) assume that that distinction makes sense. But the distinction is meaningless, it is useless, it is part of a metaphysical way of setting things up, and as such Rorty rejects it. If we really wanted to use metaphysical jargon, then we would have to say that quarks and human rights do not differ in "ontological status," and all the *objectivity* that quarks may possess is the objectivity that human rights possess, and vice versa. And this is an objectivity stemming from our discourse, from the ways in which we talk about quarks and the (different) ways in which we talk about human rights. A kind of objectivity better describable as *intersubjectivity* (this is what solidarity boils down to), something we may reach quite irrespective of whether quarks do exist independently or not. But if the only way in which quarks can be causally independent of us is that they are considered so by us—that they are "an unquestioned part of our talk"—then, in light of previous chapters, we can only expect a sharp disagreement from Putnam, since it seems that no room is left for a genuine acknowledgment of their independence. What Putnam would say is that Rorty's talk of the independence of reality is a mere *façon de parler*.

Nevertheless, Rorty believes that there are important points of contact between his own thought and Putnam's. And he lists them by quoting passages from some of Putnam's writings:

(I) ". . . elements of what we call 'language' or 'mind' *penetrate so deeply into what we call 'reality' that the very project of representing ourselves as being 'mappers' of something 'language-independent' is fatally compromised from the start* [. . .]."

(II) "[We should] accept the position we are fated to occupy in any case, the position of beings who cannot have a view of the world that does not reflect our interests and values, but who are, for all that, committed to regarding some views of the world—and, for that matter, some interests and values—as better than others."

(III) "What Quine called 'the indeterminacy of translation' should rather be viewed as the '*interest relativity of translation.*' . . . '[I]nterest relativity' contrasts with *absoluteness,* not with objectivity. It can be objective that an interpretation or an explanation is the correct one, *given* the interests which are relevant in the context."

(IV) "The heart of pragmatism, it seems to me—of James' and Dewey's pragmatism, if not of Peirce's—was the insistence on the supremacy of the agent point of view. If we find that we must take a certain point of view, use a certain 'conceptual system,' when we are engaged in practical activity, in the widest sense of 'practical activity,' then we must not simultaneously advance the claim that it is not really 'the way things are in themselves'."

(V) "To say, as [Bernard] Williams sometimes does, that convergence to one big picture is required by the very concept of knowledge is sheer dogmatism. . . . It is, indeed, the case that ethical knowledge cannot claim absoluteness; but that is because the notion of absoluteness is incoherent" (Rorty 1993: 43–44).

I have quoted the passage at length because it is a good snapshot of what can be considered points of convergence between the two philosophers—at least on paper. Point (I) we have already encountered above, noting that what is in common is the letter but not the substance of the statement, since they associate different senses to it. And the same is true for the other four statements. The reason for this dissimilarity I think is clear to the reader by now, but let us see Putnam's reaction to go into detail.

Putnam's Reaction

I think it is appropriate to begin this section by emphasizing the long and deep friendship that Rorty and Putnam managed to establish over the years. It may seem like a trite statement, but I believe that it is characteristic of great people to be able to perceive greatness in others in spite of what can sometimes come to be a strong difference of opinion. Precisely this happened to the two of them:

> He was, over many years, both a personal friend and a philosopher with whom I loved to argue. I do not think either of us ever succeeded in changing the other's mind, but I know that thinking about the issues he raised was enormously valuable for me, and I believe he would say the same about me were he alive today. [He was] a truly unique philosopher and a "gadfly" in the best tradition of the term (Putnam 2015n: 882).[16]

This friendship naturally involved Ruth Anna Putnam as well, and also for her the value of this friendship was not a veil when it came to assessing the difference between their philosophical views. Take the following brief general assessment of Rorty's thought as an example of this:

> Richard Rorty calls himself a pragmatist, but I am inclined to think that his pragmatism is profoundly different from that of, say, John Dewey. The key words in Dewey's philosophy, as I understand it, are "interaction" and "inquiry"; the key words in Rorty's recent philosophy are "conversation" and "solidarity." Not that Dewey would not approve of conversation and solidarity—both are essential to inquiry—but he would insist that what prompts the inquiry and what must be its ultimate upshot is *experience*, that is, interactions between a human organism and its environment. I have been puzzled for years why Rorty fails to note the role of experience in Dewey's thinking; the word "experience" occurs in the titles of several of Dewey's most important later books. Nor is this emphasis on experience unique to Dewey; we find it as well in the philosophies of Peirce and of James (Ruth Anna Putnam 2002: 13).

This judgment was echoed by her husband: "Fidelity to texts was not one of Dick Rorty's strengths" (Putnam 2015o: 6), "a philosopher for whom talk of a 'general progress of rationality' is, to put it mildly, problematic" (Putnam 2000a: 81), given that experience is a central element in our rational activity. But why, according to the Putnams, would experience not be given proper consideration by Rorty? As Ruth Anna Putnam reminds us, Rorty's key words are "conversation" and "solidarity." Why then is focusing on these tantamount to not adequately considering experience?

We might say that the relevance of experience resides, among other things, in justifying our statements (both empirical and non-empirical) and—when they are so justified and therefore true—in giving us the world. Now, as we have seen above Rorty conceives of justification as something "sociological": to see whether or not a statement made by a speaker S is justified (or warranted assertible, to use Dewey's term) one must see whether or not there is social solidarity with S's statement. As Rorty clarifies: "I view warrant as a sociological matter, to be ascertained by observing the reception of S's statement by his peers" (Rorty 1993: 50). To the obvious rejoinder that S's peers may be wrong and that, therefore, "whether a statement is warranted or not is independent of whether the majority of one's cultural peers would *say* it is warranted or unwarranted" (Putnam 1990a: 21),[17] Rorty gives an answer which Putnam finds "puzzling." Here is the answer:

> Well, maybe a *majority* can be wrong. But suppose everybody in the community, except for one or two dubious characters notorious for making assertions even stranger than *p,* thinks S must be a bit crazy. They think this even after patiently listening to S's defense of *p,* and after making sustained attempts to talk him out of it. Might S still be *warranted* in asserting *p*? Only if there were some way of determining warrant *sub specie aeternitatis,* some natural order of reasons that determines, quite apart from S's ability to justify *p* to those around him, whether he is justified in holding *p* (Rorty 1993: 50).

The reason why Putnam finds this puzzling is that

> it is hard to see how the sociologist, qua *sociologist,* could determine that S is warranted in asserting *p when a majority of S's cultural peers disagree.* (How does "Maybe a majority can be wrong" cohere with the claim that what is and is not a warrant for asserting something is a sociological question? Can a sociologist, qua sociologist, determine that a majority is wrong? How?—by determining that the majority contains some dubious characters? Is "dubious character" a sociological notion?) (Putnam 2000a: 84)

Putnam's conclusion is that not only is Rorty's picture of warrant deficient, not only this picture reveals that he is "a textbook example of a 'relativist'" (Putnam 2015n: 887) despite Rorty's claims to the contrary,[18] not only "nothing could have been farther from Dewey's mind than the idea that warranted assertability is just a culture-relative matter, something to be ascertained by sociologists" (Putnam 2011: 46), but his claim that warrant is a sociological notion is totally unrelated to Rorty's own practice: "when, after all, has Rorty shown the slightest interest in *sociological* description of the *actual* norms and standards current in what he calls 'our' societies?" (Putnam 2000a: 86)

However, in the last quotation from Rorty there appears a typical trait of his (in the sense that it is expressed in multiple ways and on multiple occasions in his writings) that, in Putnam's view, represents his Achilles' heel. Before evaluating it, let us see what bearing Rorty's two key words mentioned above have on the notion of the world.

We saw in the previous section that Rorty does not deny that we human beings are in contact with the world, he does not deny the existence of the world; what he denies is that we have *representational* contact with it. For him, our relation to the world is purely *causal*, and so is our experience. Putnam elucidates what is behind this thesis: it amounts to claiming that our relationship with the world is *non-semantic*, that is, not based on a relationship between language and the world that starts from a "reference" relationship between words and objects. For Rorty, it is not an individual relationship between parts of language and parts of the world, but a global, holistic relationship capable of ensuring immediate contact. In his terms:

> Our language—conceived as the web of inferential relationships between our uses of vocables—is not, on this view, something "merely human" which may hide something which "transcends human capacities." Nor can it deceive us into thinking ourselves in correspondence with something like that when we really are not. On the contrary, using those vocables is as direct as contact with reality can get (as direct as kicking rocks, e.g.). The fallacy comes in thinking that the relationship between vocable and reality has to be piecemeal (like the relation between individual kicks and individual rocks), a matter of discrete component capacities to get in touch with discrete hunks of reality (Rorty 1986: 145–146).

To which he adds:

> a distinction upon which I have insisted [is] that between "the world" (a notion which, I argued, we would be better off without) and the collection of entities such as stars, people, beavers, numbers, poems, governments, and positrons. This is the difference between something we can never be sure that we are in touch with and an assortment of things that nobody has ever been able to doubt that we are in touch with (Rorty 2015: 875).

That we are in "contact with reality" "sounds surprisingly realist" (Putnam 2015n: 887) to Putnam, and it is probably to be explained by the centrality of common sense in the thinking of the classical pragmatists—thinking that in turn was central to Rorty, of course. But, if that is the case, Putnam observes that "*to preserve our commonsense realist convictions it is not enough to preserve some set of 'realist' sentences: the interpretation you give those sentences, or, more broadly, your account of what understanding them consists in, is also important*" (Putnam 2000a: 83; emphasis original), and concludes that the interpretation Rorty puts "*upon those 'vocables' violate our deepest intuitions about what we are doing when we assert them*" (Putnam 2000a: 82; emphasis original). Indeed, it is among our pre-philosophical intuitions that when we use language we are not "merely" using language, we are not engaging in a myriad of language games that remain solely within language, since language is not a mere "web of inferential relationships between our uses of vocables" that has a general relation to a general reality. On the contrary, according to common sense our vocables have a *piecemeal* relation of reference to objects in the world.[19]

> From within our scientific and commonsense descriptions, reality is full of "discrete hunks" […]. Indeed, Rorty himself uses *kicking a rock* as an example of something that relates a particular kick to a particular rock. Is Rorty claiming that kicking the rock involves a particular rock, but describing the rock does not involve that same particular rock? How can *that* be? How can Rorty so much as use *words* to tell us that kicking a rock involves a particular rock if those very words do not relate particularly to kicks and to rocks? (Putnam 2000a: 83–84)

How could he "*say* that we are 'in touch' with people, beavers, numbers, poems, governments, and positrons if the [respective words] do not stand in any relation to those very objects; or if they do, why was 'representationalism' supposed to be a 'no-no'?" (Putnam 2015n: 887) The replacement of "talk of experience with talk of language" (Rorty 2015: 876)—that is, with nothing but conversation and solidarity—seems then to deprive Rorty of any plausible notion of reality. This, according to Putnam, is the actual bearing the two key words mentioned above have on the notion of the world.

Thus, Rorty seems to end up losing the world, so to speak—not just the world as it emerges from the implausible God's Eye View, but the world *tout court*. And one may wonder how he arrived at this surprising result, if he did: how he did not see some other more plausible path to take, once he rejected the worldview of the metaphysical realist that in the previous chapters we called "radically non-epistemic." What I termed "his typical trait" above is precisely that of not discerning any intermediate alternative to the radical realism/radical non-realism opposition. As in: either we have A, or we have non-A (roughly: either we have a non-human absolute *sub specie aeternitatis* world and a non-human absolute *sub specie aeternitatis* warrant for our statements, or we have a wholly human world and a wholly human warrant for our statements):

> We pass from the (metaphysical realist's) perception of us as being *able* to step outside of our skins to a perception of us as being *unable* to do so. We now see ourselves as forever sealed *within* our skins: confined, as it were, to *our* forms of language and thought (Conant 1994: xxvi).

It is from the conviction of the existence of this confinement that arise Rorty's anti-representationalism and the deflationary idea that truth is nothing more than a compliment. But it is difficult to understand why from the correct statement that we have no access to unconceptualized reality it should necessarily follow "that language and thought do not describe something outside themselves, even if that something can only be described by describing it (that is by employing language and thought)" (Putnam 1994g: 297). This is the step taken by Rorty and, in light of the existence of other possibilities not even considered by him, it becomes natural to wonder in the wake of Putnam why Rorty did not notice or, noticing them, did not take them into account. The diagnosis that Putnam makes of this is subtle and bold: unbeknownst to him, Rorty would be "in the grip of the picture that the Eliminative Materialism is true of the Noumenal World, even if he is debarred by the very logic of his own position from stating that belief" (Putnam 1992a: 74). It is—says Putnam—as if Rorty's analytic past reverberates in his later sharp departure from analytic philosophy:

> when he rejects a philosophical controversy, as, for example, he rejects the "realism/antirealism" controversy, or the "emotive/cognitive" controversy, his rejection is expressed in a Carnapian tone of voice—he *scorns* the controversy (Putnam 1990a: 20),

demonstrating that "he does retain strong traces of his physicalist past" (Putnam 1994g: 305). Indeed, Rorty was Carnap's student at the University of Chicago, where he received his Ph.D. in 1956, and something of Carnap's philosophical spirit must have remained with him, such as the thesis that much, if not all, of the metaphysicians' questions are nonsense. Only, "what changed with 'The World Well Lost' was that Rorty broke with the idea that sticking to scientific language is the way to avoid 'metaphysics'" (Putnam 2015n: 883).

But with this we enter somewhat slippery, psychological ground. Less psychological and more philosophical, and therefore more interesting for our purposes, is Putnam's observation about the *impossibility* which is at issue when one argues—in line with both Rorty and Putnam—that it is impossible to stand outside our thought and language, and compare thought and language, on the one hand, with the world, on the other. What both philosophers maintain is that it is *unintelligible* to say such a thing; but

> if we agree that it is *unintelligible* to say, "We sometimes succeed in comparing our language and thought with reality as it is in itself," then we should realize that it is also unintelligible to say, "It is *impossible* to stand outside and compare our thought and language with the world" (Putnam 1994g: 299).

However, it is precisely the latter that seems to be the conclusion Rorty had reached, a conclusion that—in the absence of a supporting argument—remains unintelligible because unwarranted. It is unwarranted in the sense that it is not clear why it should be the only alternative left in the field after the recognition of the unintelligibility of radically non-epistemic metaphysical realism. According to Putnam, what Rorty failed to appreciate is that the negation of something unintelligible does not in itself configure an impossibility.[20] After all, it *is* possible to stand outside and compare our thought and language with the world: it's all a question of properly characterizing this "standing outside" and the related relationship to the world. This is not an impossibility. The lack of an "absolute" guarantee that words represent things outside of themselves does not necessarily carry with it the impossibility of representation *tout court*. Therefore,

> Failing to inquire into the character of the unintelligibility which vitiates metaphysical realism, Rorty remains blind to the way in which his own rejection of metaphysical realism partakes of the same unintelligibility (Putnam 1994g: 300).

All of this led Rorty to turn his back mistakenly both on the *platitudinous* idea that language can represent something which is itself outside of language, and on commonsense realism, "the realism that says that mountains and stars are not created by language and thought, and are not parts of language and thought, and yet can be described by language and thought" (Putnam 1994g: 303).[21]

Thus, there emerges in Putnam's exchange with Rorty over more than four decades the scope and significance that the classical pragmatists had for him. The important pragmatist legacy concerning epistemological, metaphysical, and moral issues was a very important source of inspiration for Putnam, who drew important lessons from it by combining it in very personal ways with his early training. In particular, the central philosophical aim of the classical pragmatists—"the aim of producing a philosophy which would do justice to our respect for science, and for fact in general, *as well as* our humanistic values and concerns (and, in James's case, our religious feelings)" (Putnam 2011: 47)—was a very fruitful way of endowing his particular version of naturalism, *liberal naturalism*, with content and showing how to make philosophy matter for life. Above all, it opened a way to show the imperishable value of philosophy, in spite of the denigrators who down the ages have never been lacking. In this regard Putnam was fond of quoting the following passage from Étienne-Henri Gilson: "the first law to be inferred from philosophical experience is: *Philosophy always buries its undertakers*" (Gilson 1937: 306). And this seems to me an excellent closing for a book dedicated to a thinker whose full impact on philosophy it is for future generations to appreciate.

NOTES

1. Cf.: "To tell the truth, I have never thought that classifying philosophy as 'analytic' or 'continental' is a good thing" (Putnam 2004a: 111). This is a distinction that is no longer useful because according to Putnam both labels have lost their original meaning (assuming they had any). "I think the term Continental philosophy is no longer a good one, because national differences have reappeared. The difference now between German philosophy and French philosophy is so great that to use the term Continental philosophy is no longer useful." And, as for analytic philosophy, "I believe that analytic philosophy started with respect for argument. The problem is that after a while all philosophy had to be argument, and people didn't know what to argue about. Therefore, imaginary objects emerged: possible worlds are just as real as imaginary ones. […] I don't criticize analytic philosophy for being metaphysical, I am not an anti-metaphysics militant. My problem with analytic philosophy is that it is empty. All philosophy does not have to be argument, and all arguments do not have to be in the analytic style" (both quotes are from Borradori 1994: Chapter 3: "Between the New Left and Judaism: Hilary Putnam," 59 and 69).

2. For the same reason Putnam does not consider the figure of the "disciple" a good thing: "I've learned from a number of my teachers, but without even becoming a disciple of any of them. I don't believe in disciples, by the way, and I haven't produced any permanent disciples myself. A few of my former students tried to be my disciples, but they decided they didn't agree with me when I gave up whichever position I had defended when they were my students" (Putnam 1997: 149–150). It should be made clear, however, that this criticism of movements is not the same as disavowing their importance: simply, they need to be considered in the right light. For example, with regard to pragmatism he argued the following: "I am sure pragmatism has a future provided it does not degenerate into the idea that some group of past thinkers has all the answers. It is dangerous to say 'The best is behind us, you just have to go back to Dewey, to James, or to Peirce.' That's not the way to keep pragmatism alive. I believe both Rorty and I agreed on that" (Putnam 2015o: 9).

3. Cf.: "what makes me say I am not a 'card-carrying' [pragmatist] is that I have *never* agreed with the several pragmatist 'theories of truth' nor with the several grand metaphysical theories that James, Dewey, and Peirce proposed, although I certainly learned a great deal from reading them and from giving courses and seminars on their

work" (Putnam 2015b: 91). Cf. also: "the antirealist and/or verificationist views of the classical pragmatists are no part of what I admire in their philosophies" (Putnam 2015m: 773), Putnam (2012c): 70, and Rorty, Putnam, Conant, & Helfrich (2004): 74. For interesting analyses of James's conception of truth and Putnam's, cf. Cormier (2015) and Jackman (2021).

4. Cf. also: "although some have tried to pin the label 'pragmatist' on me, my attitude toward the pragmatists is similar to my attitude toward Wittgenstein. I believe we can learn from them without accepting every one of the views attributed to them (rightly or wrongly). If there is a similarity between my views and Wittgenstein's, or my views and Dewey's, or my views and James's, it stems from the fact that we are all, albeit in different ways, non-reductive and non-scientistic naturalists. But a similarity is not an identity" (Putnam 2008f: 337).

5. "I paid little attention to this at the time (I was much more interested in Freud, Kierkegaard, Marx—until I read A. J. Ayer's *Language, Truth and Logic* and became briefly 'converted' to Ayer's view)," Putnam 2002b: 19–20.

6. For one of James's roles in shaping Putnam's thinking, cf. Putnam (2016a): 224ff. For a useful discussion of Putnam's relationship with James, Dewey, C. I. Lewis, as well as for his ties to certain aspects of Stanley Cavell's thought, cf. Goodman (2008). For the combination of pragmatism and pluralism in Putnam, cf. Goodman (2013).

7. For an analysis of Putnam's interpretation of Dewey's epistemology and metaphilosophy, cf. Hildebrand (2000).

8. The other names mentioned by Conant are: Aristotle, Kant, Wittgenstein, Jürgen Habermas, Richard Rorty, Bernard Williams, Stanley Cavell, Cora Diamond, John McDowell, and Charles Travis.

9. For a discussion of the extension of the indispensability argument to the sphere of ethics in the broader context of the exchange between Putnam and Habermas, cf. Gil Martín & Vega Encabo (2008).

10. Putnam takes Penelope Maddy's work as an example here.

11. The primary reference here is to John Mackie.

12. Here is another claim by Rorty to the same effect: "There is no way, as Wittgenstein has said, to come between language and its object, to divide the giraffe in itself from our ways of talking about giraffes" (Rorty 1999: xxvii).

13. Rorty does not deny that there is truth: "'There is no truth.' What could that mean? Why should anybody say it? Actually, almost nobody [...] does say it" (Rorty 1998a: 1). Cf. also: "To say that we should drop the idea of truth as out there waiting to be discovered is not to say that we have discovered that, out there, there is no truth" (Rorty 1989: 8).

14. Cf.: "To say that Freud's vocabulary gets at the truth about human nature, or Newton's at the truth about the heavens, is not an explanation of anything. It is just an empty compliment—one traditionally paid to writers whose novel jargon we have found useful" (Rorty 1989: 8).

15. The *cautionary* use of the word "true" is "the use found in sentences such as 'Your arguments satisfy all our contemporary norms and standards, and I can think of

nothing to say against your claim, but still, what you say might not be true'" (Rorty 1993: 60).

16. Cf. also: "We were not enemies, we loved each other, but we differed very much, and differences were stimulating. I think I clarified my own positions in my head by seeing where I disagreed with Rorty" (Putnam 2015o: 6).

17. That warrant is independent of majority opinion is according to Putnam a *central part* of our picture of warrant: "To say that whether or not it is warranted in a given problematical situation to accept a given judgment is independent of whether a majority of one's peers would *agree* that it is warranted in that situation is just to show that one has the concept of warrant" (Putnam 1990a: 22).

18. For a useful discussion of Rorty and Putnam on relativism, cf. Case (1995) and Hartz (1991).

19. To be fair to Rorty, however, we must point out that he believed that common sense is too imbued with the picture of the "mirror of nature" to be of any philosophical use; for this reason he thinks it is not totally reliable and therefore should be reformed: "If contemporary intuitions are to decide the matter, 'realism' and representationalism will always win, and the pragmatists' quietism will seem intellectually irresponsible. So pragmatists should not submit to their judgment. Instead, they should see themselves as working at the interface between the common sense of their community, a common sense much influenced by Greek metaphysics and by monotheism, and the startingly counterintuitive self-image sketched by Darwin, and partially filled in by Dewey. They should see themselves as involved in a long-term attempt to change the rhetoric, the common sense, and the self-image of their community" (Rorty 1995: 41).

20. On this issue cf. Conant (1992).

21. For a different assessment of the relationship between Putnam and Rorty, cf. Cormier (2006) and Schwartz (2021). For a discussion of the pragmatist positions of the two philosophers that also involves Robert Brandom's pragmatism, cf. Baghramian (2008b).

Epilogue
The End of a Journey

I would like to close the book on a personal note. We have seen how Putnam used to refer to his decade-long reflection on naturalism and realism as a "long journey" that passed through many different topics, from natural science to mathematics, language, ethics, and so on—and I and the readers who have followed the chapters of this book have done much the same. From the early analyses of the concepts of analyticity and necessity to the considerations on the cognitive value of the human sciences, passing, among other things, through the evaluation of the dynamics of scientific theories and the semantic relationship that connects language to the world, a philosophical thought in the making has developed before our eyes. Although other places could have been visited, the journey can end here. I usually ask students who have completed their dissertation to add a conclusion in which they set out what they have gained from writing the dissertation itself, what they have at the end of the work that they didn't have at the beginning. I think the same question can be asked of me now—and I would answer as follows.

First of all, I have gained an impression of a very rare thinker, especially in our times: completely immersed in his work as a teacher and scholar, with a huge production of papers on a vast number of topics, several of which are still unpublished—he used to say he was a very quick writer—always ready to admit with a great deal of intellectual honesty that he had changed his opinion on matters of fundamental importance, endowed with an "infallible instinct for what, in the unsurveyable diversity of contemporary discussions, is genuinely *significant*, combined with his ability to arrange a confrontation with the issues in a fashion that consistently promises to advance our thinking in some new direction" (Stegmüller 1987: 345; Engl. transl. by J. Conant; see Conant 1990: xxxix)—and in this comparable with few figures in the tradition of Western culture. John Passmore famously likened him to Bertrand Russell in terms of breadth of philosophical scope, critical and self-critical sense, and dedication to the work: "Putnam shares Russell's capacity for changing his mind as a result of learning from his contemporaries" (Passmore 1988: 104). This is a comparison I have always found interesting, but if I were asked to make such a comparison, I would rather think of an Italian philosopher who lived at the turn of the sixteenth and seventeenth centuries in southern Italy, Tommaso Campanella, who was deeply and assiduously interested in metaphysics, epistemology, natural sciences, ethics, politics, and religion, and who constantly reshaped his own philosophical theses in light of novel considerations (cf. Ernst 2010). Moreover, an abiding interest in both theoretical thinking and how to make philosophy matter for life—both individually and collectively—was what guided the Italian philosopher's activity. Apart from torture and imprisonment by the Inquisition, which Putnam fortunately never underwent, a striking similarity.

It is worth noting that Passmore's remark implicitly highlights the way of doing philosophy that we have been able to appreciate since the first chapter of this book: a constant confrontation with fellow philosophers, so much so that very often the cue to begin writing a paper is the desire to discuss a discordant point of view. That is especially why,

> [n]ot infrequently, it was the stimulus of other thinkers that precipitated his (in) famous changes of mind. Putnam never considered his views finally settled and constantly tried to challenge them, very often with the help of the views of agonistic philosophers. Philosophy, as he envisions it, is a collective activity conducted in a fallibilistic and democratic spirit (De Caro & Macarthur 2022c: 4).

The real significance of such a mode of doing philosophy is surely something I have learned from going through Putnam's writings to write this book. It's a method that represents philosophy in one of its highest aspirations, if it is true that it is mainly through exchange with others that we can understand more deeply problems that affect everyone. I would also go so far as to claim that this method has a sort of moral flavor to it: not only does it show interest in the other's point of view, but it also helps the interlocutor to clearly see the main points of one's own position on a given topic—weaker points included. An all-round openness to the other's point of view. Moreover, I think there is a connection between the conception of philosophy as dialogue and Putnam's changes of mind referred to in the last quotation—albeit modeling one's philosophical reflection on dialogue about a given topic cannot be the main reason, much less the only reason, for changing one's mind on that topic. This connection emerges, for example, in one of the few places where

Putnam addresses the issue of his changes of mind. After praising Carnap for the same willingness to change his mind in cases where he realized it was right to do so, he goes on to maintain that,

[a]lthough I do not now agree with Carnap's doctrines of any particular period, for me Carnap is still the outstanding example of a human being who puts the search for truth higher than personal vanity. A philosopher's job is not to produce a view X and then, if possible, to become universally known as "Mr. View X" or "Ms. View X." If philosophical investigations (a phrase made famous by another philosopher who "changed his mind") contribute to the thousands-of-years-old *dialogue* which is philosophy, if they deepen our understanding of the riddles we refer to as "philosophical problems," then the philosopher who conducts those investigations is doing the job right. Philosophy is not a subject that eventuates in final solutions (Putnam 1988: xii; my italics).

In a sense, it is the full awareness that philosophy does not eventuate in final solutions that kept Putnam's journey in motion, and me with him. Still, at the end of this journey the realization remains that I was not able to deal with *everything* Putnam covered. Accounts of Putnam's interpretation of quantum mechanics, his aesthetical ideas, his interpretation of Wittgenstein, and his conception of free will have been left out. Probably, as I said in the Introduction, to write a comprehensive book on Putnam you would need a *team* of scholars with different expertises—or you would need Putnam himself. This—to my astonishment—may be considered something that I know now and didn't know when I started writing the book.

Perhaps however—and this is my final consideration—the most important of the realizations that came to me from writing the book is how central Putnam's conception of the *synthetic a priori* is—the notion of quasi-necessary statements that we encountered in the first chapter. I think it is central not only to Putnam's philosophy, but also to contemporary reflection in general. The idea was in some measure anticipated by the later Wittgenstein and by Norwood Russell Hanson, and it was taken into account by Quine and developed by Thomas Kuhn in terms of the notion of framework principles of a paradigm—principles that are immune to empirical verification in periods of the so-called "normal science." But it seems to me that Putnam's notion is preferable because it is not entangled with the antirealist outcomes that seem inevitable in the (early) Kuhnian position; it is more precise than Quine's notion of "centrally located within the total network" statements (Quine 1951: 44); and, although analogous to Wittgenstein's notion of *hinge propositions*, it is discussed by Putnam in a wider variety of cases and more clearly characterized as something that "cannot be overthrown by experiments and observations *alone*, but only by alternative theories" (Putnam 1974: 259). It represents a conception dense of consequences that Putnam himself wasn't initially fully aware of.

For all this, what I have gained at the end of writing this book is the figure of a much greater thinker than the image I had of him at first.

Bibliography

Achinstein, P. & Barker, S. (eds.) (1969) *The Legacy of Logical Positivism: Studies in the Philosophy of Science*, Baltimore: Johns Hopkins Press.

Alai, M. (1994) "Brains in Vat and their Minds: A Wrong Impossibility Proof," *European Review of Philosophy* 1: 3–18.

Alai, M. (2012) "Levin and Ghins on the 'No Miracle' Argument and Naturalism," *European Journal for Philosophy of Science* 2: 85–110.

Alai, M. (2014) "Novel Predictions and the No Miracle Argument," *Erkenntnis* 79: 297–326.

Alai, M. (2017) "Resisting the Historical Objections to Realism: Is Doppelt's a Viable Solution?," *Synthese* 194: 3267–3290.

Alai, M. (2020) "Scientific Realism, Metaphysical Antirealism and the No Miracle Arguments," *Foundations of Science*, DOI 10.1007/s10699-020-09691-z.

Audi, R. (ed.) (2015) *The Cambridge Dictionary of Philosophy*, 3rd edition, Cambridge: Cambridge University Press.

Austin, J. (1946) "Other Minds," *Proceedings of the Aristotelian Society*, Suppl. 20: 148–187; repr. in Austin (1961): 76–116.

Austin, J. (1961) *Philosophical Papers*, Oxford: Clarendon Press.

Austin, J. (1962) *Sense and Sensibilia*, Oxford: Clarendon Press.

Auxier, R. E. & Hahn, L. E. (eds.) (2007) *The Philosophy of Michael Dummett*, Chicago: Open Court.

Auxier, R. E., Anderson, D. R. & Hahn, L. E. (eds.) (2015) *The Philosophy of Hilary Putnam*, Chicago: Open Court.

Baghramian, M. (2008a) "'From Realism Back to Realism': Putnam's Long Journey," *Philosophical Topics* 36: 17–35.

Baghramian, M. (2008b) "Three Pragmatisms: Putnam, Rorty, and Brandom," in Rivas Monroy, Cancela Silva & Martínez Vidal (2008): 83–101.

Baghramian, M. (ed.) (2013) *Reading Putnam*, London: Routledge.

Bays, T. (2001) "On Putnam and His Models," *The Journal of Philosophy* 98: 331–350.

Benacerraf, P. (1965) "What Numbers Could Not Be," *The Philosophical Review* 74: 47–73; repr. in Benacerraf & Putnam (1983): 272–294.

Benacerraf, P. & Putnam, H. (eds.) (1983) *Philosophy of Mathematics: Selected Readings*, 2nd edition, Cambridge: Cambridge University Press.

Ben-Menahem, Y. (ed.) (2005) *Hilary Putnam*, Cambridge: Cambridge University Press.

Ben-Menahem, Y. (2022) "Natural Laws and Human Language," in Conant & Chakraborty (2022): 289–308.

Berlin, I. (2000) *The Power of Ideas*, 2nd edition, Princeton: Princeton University Press, 2013.

Bernecker, S. (2016) "Extended Minds in Vats," in Goldberg (2016): 54–72.

Bernstein, R. J. (2005) "The Pragmatic Turn: The Entanglement of Fact and Value," in Ben-Menahem (2005): 251–265.

Berti, E. (2001) "Multiplicity and Unity of Being in Aristotle," *Proceedings of the Aristotelian Society* 101: 185–207.

Biletzki, A. & Matar, A. (eds.) (1998) *The Story of Analytic Philosophy: Plot and Heroes*, New York: Routledge.

Block, N. (ed.) (1980) *Readings in the Philosophy of Psychology*, Cambridge, MA: Harvard University Press.

Borradori, G. (1994) *The American Philosopher: Conversation with Quine, Davidson, Putnam, Nozick, Danto, Rorty, Cavell, MacIntyre, and Kuhn*, Chicago: University of Chicago Press; originally published in Italian, Roma-Bari: Laterza, 1991.

Boyd, R. (1980) "Materialism without Reductionism: What Physicalism Does Not Entail," in Block (1980): 67–106.

Boyd, R. (1991) "On the Current Status of Scientific Realism," in Boyd, Gasper & Trout (1991): 195–222.

Boyd, R., Gasper, P. & Trout, J. D. (eds.) (1991) *The Philosophy of Science*, Cambridge, MA: MIT Press.

Braddon-Mitchell, D. & Nola, R. (eds.) (2009) *Conceptual Analysis and Philosophical Naturalism*, Cambridge, MA: MIT Press.

Brandom, R. B. (ed.) (2000) *Rorty and His Critics*, Malden: Blackwell.

Brueckner, A. (1986) "Brains in a Vat," *Journal of Philosophy* 83: 148–167.

Brueckner, A. & Ebbs, G. (2012) *Debating Self-Knowledge*, Cambridge: Cambridge University Press.

Brunsveld, N. (2017) *The Many Faces of Religious Truth: Hilary Putnam's Pragmatic Pluralism on Religion*, Leuven: Peeters.

Buber, M. (1923) *Ich und Du*, Leipzig: Insel; Engl. transl. *I and Thou*, W. Kaufmann (ed.), New York: Scribner, 1970.

Buechner, J. (2008) *Gödel, Putnam, and Functionalism: A New Reading of Representation and Reality*, Cambridge, MA: MIT Press.

Burge, T. (1979) "Individualism and the Mental," in French, Uehling & Wettstein (1979): 73–121; partially repr. in Pessin & Goldberg (1996): 125–141.

Burgess, J. P. (2018) "Putnam on Foundations: Models, Modals, Muddles," in Hellman & Cook (2018): 129–143.

Butler, R. (ed.) (1963) *Analytical Philosophy, Second Series*, Oxford: Basil Blackwell.

Button, T. (2013) *The Limits of Realism*, Oxford: Oxford University Press.

Capitan, W. H. & Merrill, D. D. (eds.) (1967) *Art, Mind and Religion*, Pittsburgh: University of Pittsburgh Press.

Carnap, R. (1928) *Der logische Aufbau der Welt*, Berlin: Weltkreis.

Carnap, R. (1936-37) "Testability and Meaning," *Philosophy of Science* 3 (1936): 420–468, and 4 (1937): 1–40; repr. in Feigl & Brodbeck (1953): 47–92.

Carnap, R. (1947) *Meaning and Necessity*, Chicago: University of Chicago Press.

Carnap, R. (1950) "Empiricism, Semantics and Ontology," *Revue internationale de philosophie* 11: 20–40.

Carnap, R. (1956) "The Methodological Character of Theoretical Concepts," in Feigl & Scriven (1956): 38–75.

Carr, D. & Cheung, C.-F. (eds.) (2004) *Time, Space, and Culture*, Dordrecht: Springer.

Case, J. (1995) "Rorty and Putnam: Separate and Unequal," *Southern Journal of Philosophy* 33: 169–184.

Case, J. (1997) "On the Right Idea of a Conceptual Scheme," *Southern Journal of Philosophy* 35: 1–18.

Castañeda, H.-N. (ed.) (1967) *Intentionality, Minds, and Perception: Discussions on Contemporary Philosophy: A Symposium*, Detroit: Wayne State University Press.

Cauman, L. S., Levi, I., Parsons, C. & Schwartz, R. (1983) *How Many Questions? Essays in Honor of Sidney Morgenbesser*, Indianapolis: Hackett.

Cavell, S. (1979) *The Claim of Reason: Wittgenstein, Skepticism, Morality, and Tragedy*, Oxford: Oxford University Press.

Chakraborty, S. (2019) "The Fact/Value Dichotomy: Revisiting Putnam and Habermas," *Philosophia* 47: 369–386.

Chakraborty, S. (2020) *The Labyrinth of Mind and World: Beyond Internalism–Externalism*, New York: Routledge.

Chakraborty, S. (2022) "Language, Meaning, and Context Sensitivity: Confronting a 'Moving-Target'," in Conant & Chakraborty (2022): 89–106.

Clark, K. J. (ed.) (2016) *The Blackwell Companion to Naturalism*, Malden: Blackwell.

Clark, P. & Hale, B. (eds.) (1994) *Reading Putnam*, Oxford: Basil Blackwell.

Coliva, A. (2015) *Extended Rationality: A Hinge Epistemology*, Basingstoke: Palgrave Macmillan.

Colodny, R. G. (ed.) (1962) *Frontiers of Science and Philosophy*, Pittsburgh: University of Pittsburgh Press.

Colodny, R. G. (ed.) (1965) *Beyond the Edge of Certainty: Essays in Contemporary Science and Philosophy*, Englewood Cliffs: Prentice-Hall.

Compagnon, A. (1998) *Le démon de la théorie: Littérature et sens commun*, Paris: Seuil.

Conant, J. (1990) "Introduction," in Putnam (1990a): xv–lxxiv.

Conant, J. (1992) "The Search for Logically Alien Thought: Descartes, Kant, Frege, and the *Tractatus*," in Hill (1992): 115–180.

Conant, J. (1994) "Introduction," in Putnam (1994a): xi–lxxvi.

Conant, J. (2022) "An Introduction to Hilary Putnam," in Conant & Chakraborty (2022): 1–46.

Conant, J. & Chakraborty, S. (eds.) (2022) *Engaging Putnam*, Berlin: de Gruyter.

Conant, J. & Zeglen, U. M. (eds.) (2002) *Hilary Putnam: Pragmatism and Realism*, Abingdon: Routledge.

Cormier, H. J. (2006) "Hilary Putnam," in Shook & Margolis (2006): 108–119.

Cormier, H. J. (2015) "What Is the Use of Calling Putnam a Pragmatist?," in Auxier, Anderson & Hahn (2015): 801–819.

Crețu, A.-M. & Massimi, M. (eds.) (2020) *Knowledge from a Human Point of View*, Cham: Springer.

Dancy, J. (2017) "Moral Particularism," in *The Stanford Encyclopedia of Philosophy* (Winter 2017 Edition), E. N. Zalta (ed.), https://plato.stanford.edu/entries/moral-particularism.

Davidson, D. (1984) *Inquiries into Truth and Interpretation*, Oxford: Clarendon Press.

Davidson, D. (1995) "The Objectivity of Values," in Davidson (2004): 39–52.

Davidson, D. (2004) *Problems of Rationality*, Oxford: Clarendon Press.

Davies, R. (2004) "The Demon and the Scientist," *Epistemologia* 27: 299–318.

Davis, M. (2018) "Pragmatic Platonism: Mathematics and the Infinite," in Hellman & Cook (2018): 145–159.

De Caro, M. (2015) "Realism, Common Sense, and Science," *The Monist* 98: 197–214.

De Caro, M. (2016) "Putnam's Philosophy and Metaphilosophy," in Putnam (2016a): 1–18.

De Caro, M. (2018) "Putnam on the Mind-Body Problem," *Belgrade Philosophical Annual* 31: 155–163.

De Caro, M. (2019) "Common-sense and Naturalism," in Giladi (2019): 184–207.

De Caro, M. (2020) "Hilary Putnam on Perspectivism and Naturalism," in Crețu & Massimi (2020): 57–70.

De Caro, M. (2022) "Davidson and Putnam on the Antinomy of Free Will," in Conant & Chakraborty (2022): 249–261.

De Caro, M. (forthcoming) "Putnam's Liberal Naturalism," in Frauchiger (forthcoming).

De Caro, M. & Floyd, J. (eds.) (2020) "Liberal Naturalism: The Legacy of Hilary Putnam," *The Monist* 103: 357–492.

De Caro, M. & Macarthur, D. (eds.) (2004) *Naturalism in Question*, Cambridge, MA: Harvard University Press.

De Caro, M. & Macarthur, D. (eds.) (2010) *Naturalism and Normativity*, New York: Columbia University Press.

De Caro, M. & Macarthur, D. (2012) "Hilary Putnam: Artisanal Polymath of Philosophy," in Putnam (2012a): 1–35.

De Caro, M. & Macarthur, D. (2015) "Liberal Naturalism," in Audi (2015): 595–596.

De Caro, M. & Macarthur, D. (eds.) (2022a) *The Routledge Handbook of Liberal Naturalism*, Abingdon: Routledge.

De Caro, M. & Macarthur, D. (2022b) "Introduction," in De Caro & Macarthur (2022a): 1–4.

De Caro, M. & Macarthur, D. (2022c) "Hilary Putnam: Dialogical Philosopher," in Putnam (2022): 1–6.

De Caro, M. & Putnam, H. (2020) "Free Will and Quantum Mechanics," *The Monist* 103: 415–426.

de Gaynesford, M. (2006) *Hilary Putnam*, Chesham: Acumen.

de Gaynesford, M. (2022) "Balance in *The Golden Bowl*: Attuning Philosophy and Literary Criticism," in Conant & Chakraborty (2022): 309–330.

Dell'Utri, M. (1990) "Choosing Conceptions of Realism: The Case of the Brains in a Vat," *Mind* 99: 79–90.

Dell'Utri, M. (2008) "The Threat of Cultural Relativism: Hilary Putnam and the Antidote of Fallibilism," *European Journal of Analytic Philosophy* 4 (2): 75–86.

Dell'Utri, M. (2016) "Putnam's Conception of Truth," *European Journal of Analytic Philosophy* 12: 5–22.

Dell'Utri, M. (2017) "'Metaphysics Without Ethics is Blind': The Legacy of Hilary Putnam," *Journal for General Philosophy of Science* 48: 501–515.

Dell'Utri, M. (2021) "Hilary Putnam: Embodying Philosophy," *Share* 15: 26–27.

Dell'Utri, M. (2022) "Putnam and Liberal Naturalism," in De Caro & Macarthur (2022): 455–463.

De Mol, L. (2018) "Turing Machines," in *The Stanford Encyclopedia of Philosophy*, E. N. Zalta (ed.), https://plato.stanford.edu/entries/turing-machine/.

Dewey, J. (1925) *Experience and Nature*, 2nd edition, London: Allen & Unwin, 1929.

do Carmo, J. S. (ed.) (2015) *A Companion to Naturalism*, Pelotas: Nepfil.

Douven, I. (1999) "Putnam's Model-Theoretic Argument Reconstructed," *The Journal of Philosophy* 96: 479–490.

Dummett, M. (1969) "The Reality of the Past," in Dummett (1978): 358–374.

Dummett, M. (1975) "What is a Theory of Meaning?," in Guttenplan (1975): 97–138; repr. in Dummett (1993): 1–33.

Dummett, M. (1976) "What is a Theory of Meaning? (II)," in Evans & McDowell (1976): 67–137; repr. in Dummett (1993): 34–93.

Dummett, M. (1978) *Truth and Other Enigmas*, Cambridge, MA: Harvard University Press.

Dummett, M. (1991) *The Logical Basis of Metaphysics*, Cambridge, MA: Harvard University Press.

Dummett, M. (1993) *The Seas of Language*, Oxford: Clarendon Press.

Dummett, M. (1998) "Truth from the Constructivist Standpoint," *Theoria* 64: 122–138.

Dummett, M. (2004) *Truth and the Past*, New York: Columbia University Press.

Dummett, M. (2007) "Reply to Hilary Putnam," in Auxier & Hahn (2007): 168–184.

Dümont, J. (1999) "Putnam's Model-Theoretic Argument: A Detailed Reconstruction," *Journal for General Philosophy of Science* 30: 341–364.

Eatwell, J., Milgate, M. & Newman, P. (eds.) (1987) *The New Palgrave: A Dictionary of Economics*, Volume 3, London: Macmillan.

Ebbs, G. (1992) "Realism and Rational Inquiry," in Hill (1992): 1–33.

Ebbs, G. (2012) "Skepticism, Objectivity, and Brains in Vats," in Brueckner & Ebbs (2012): 28–54; orig. 1992.

Ebbs, G. (2016a) "Putnam on Methods of Inquiry," *Argumenta* 2 (1): 157–161.

Ebbs, G. (2016b) "How to Think about Whether We Are Brains in Vats," in Goldberg (2016): 27–36.

Ebbs, G. (2017) *Carnap, Quine, and Putnam on Methods of Inquiry*, Cambridge: Cambridge University Press.

Ebbs, G. (2020) "Hilary Putnam's Liberal Naturalism about Language Use, Reference, and Truth," *The Monist* 103: 357–369.

Ebbs, G. (2022) "Putnam on Trans-Theoretical Terms and Contextual Apriority," in Conant & Chakraborty (2022): 131–155.

Elgin, Z. C. (1996) "The Relativity of Fact and the Objectivity of Value," *The Harvard Review of Philosophy* 6 (1): 4–15.

Elgin, Z. C. (2007) "The Fusion of Fact and Value," *Iride* 20: 83–101.

Ernst, G. (2010) *Tommaso Campanella: The Book and the Body of Nature*, D. L. Marshall (transl.), Dordrecht: Springer.

Feigl, H. & Brodbeck, M. (eds.) (1953) *Readings in the Philosophy of Science*, New York: Appleton-Century-Crofts.

Feigl, H. & Scriven, M. (eds.) (1956) *The Foundations of Science and the Concepts of Psychology and Psychoanalysis*, *Minnesota Studies in the Philosophy of Science*, Volume 1, Minneapolis: University of Minnesota Press.

Ferrari, F. (2022) *Truth and Norms: Normative Alethic Pluralism and Evaluative Disagreements*, Lanham: Lexington Books.

Fitch, F. B. (1963) "A Logical Analysis of Some Value Concepts," *The Journal of Symbolic Logic* 28: 135–142.

Floyd, J. (2005) "Putnam's 'The Meaning of "Meaning"': Externalism in Historical Context," in Ben-Menahem (2005): 17–52.

Floyd, J. (2020) "Aspects of the Real Numbers: Putnam, Wittgenstein, and Nonextensionalism," *The Monist* 103: 427–441.

Frauchiger, M. (ed.) (forthcoming) *Mind and Meaning: Themes from Putnam*, Berlin: De Gruyter.

Frege, G. (1892) "Über Sinn und Bedeutung," *Zeitschrift für Philosophie und philosophische Kritik* 100: 25–50; translated as "On Sense and Reference," by M. Black, in Geach & Black (1960): 56–78.

Frege, G. (1918-19) "Der Gedanke: Eine logische Untersuchung," *Beiträge zur Philosophie des deutschen Idealismus* 1: 58–77; translated as "Thoughts," by P. Geach & R. H. Stoothoff, in Frege (1984): 351–372.

Frege, G. (1984) *Collected Papers in Mathematics, Logic, and Philosophy*, ed. B. McGuinness, Oxford: Basil Blackwell.

French, P., Uehling, T. & Wettstein, H. (eds.) (1979) *Studies in Metaphysics*, Minneapolis: University of Minnesota Press.

French, P., Uehling, T. & Wettstein, H. (eds.) (1984) *Causation and Causal Theories*, Minneapolis: University of Minnesota Press.

French, P., Uehling, T. & Wettstein, H. (eds.) (1994) *Philosophical Naturalism*, Notre Dame: University of Notre Dame Press.

Garavaso, P. (2013) "Hilary Putnam's Consistency Objection against Wittgenstein's Conventionalism in Mathematics," *Philosophia Mathematica* 21 (3): 1–18.

García-Carpintero, M. (1996) "The Model-Theoretic Argument: Another Turn of the Screw," *Erkenntnis* 44: 305–316.

Geach, P. & Black, M. (eds.) (1960) *Translation from the Philosophical Writings of Gottlob Frege*, Oxford: Basil Blackwell.

Geertz, C. (1973a) "Thick Description: Toward an Interpretative Theory of Culture," in Geertz (1973b): 3–30.

Geertz, C. (1973b) *The Interpretation of Cultures: Selected Essays*, New York: Basic Books.

George, A. (ed.) (1989) *Reflections on Chomsky*, Oxford: Oxford University Press.

Giladi, P. (ed.) (2019) *Responses to Naturalism: Critical Perspectives from Idealism and Pragmatism*, New York: Routledge.

Gil Martín, F. J. & Vega Encabo, J. (2008) "Truth and Moral Objectivity: Procedural Realism in Putnam's Pragmatism," in Rivas Monroy, Cancela Silva & Martínez Vidal (2008): 265–285.

Gilson, É.-H. (1937) *The Unity of Philosophical Experience*, New York: Scribner's Sons, 1950.

Goldberg, S. C. (ed.) (2016) *The Brain in a Vat*, Cambridge: Cambridge University Press.

Goodman, R. B. (2008) "Some Sources of Putnam's Pragmatism," in Rivas Monroy, Cancela Silva & Martínez Vidal (2008): 125–140.

Goodman, R. B. (2013) "Some Sources of Putnam's Pluralism," in Baghramian (2013): 205–218.

Grice, H. P. & Strawson, P. F. (1956) "In Defense of a Dogma," *The Philosophical Review* 65: 141–158.

Gunderson, K. (ed.) (1975) *Language, Mind and Knowledge, Minnesota Studies in the Philosophy of Science*, Volume 7, Minneapolis: University of Minnesota Press.

Gustafsson, M. & Hertzberg, L. (eds.) (2002) *The Practice of Language*, Dordrecht: Springer.

Guttenplan, S. (ed.) (1975) *Mind and Language*, Oxford: Clarendon Press.

Guttenplan, S. (ed.) (1994) *A Companion to the Philosophy of Mind*, Oxford: Blackwell.

Hadot, P. (1981) *Exercices spirituels et philosophie antique*, Paris: Études augustiniennes; Engl. transl. *Philosophy as a Way of Life*, Oxford: Blackwell, 1995.

Hahn, L. E. & Schilpp, P. A. (eds.) (1986) *The Philosophy of W. V. Quine*, La Salle: Open Court.

Haldane, J. (2015) "Philosophy, Causality, and God," in Auxier, Anderson & Hahn (2015): 683–701.

Haldane, J. & Wright, C. (eds.) (1993) *Reality, Representation, and Projection*, Oxford: Oxford University Press.

Hale, B. & Wright, C. (eds.) (1997a) *A Companion to the Philosophy of Language*, Oxford: Blackwell.

Hale, B. & Wright, C. (1997b) "Putnam's Model-Theoretic Argument against Metaphysical Realism," in Hale & Wright (1997a): 427–457.

Harman, G. & Davidson, D. (eds.) (1972) *Semantics of Natural Language*, Dordrecht: Reidel.

Hartz, C. G. (1991) "What Putnam Should Have Said: An Alternative Reply to Rorty," *Erkenntnis* 34 (3): 287–295.

Heidegger, M. (1996) *Being and Time*, New York: State University of New York Press.

Hellman, G. (1989) *Mathematics without Numbers: Towards a Modal-Structural Interpretation*, Oxford: Oxford University Press.

Hellman, G. (2015) "Infinite Possibilities and Possibilities of Infinity," in Auxier, Anderson & Hahn (2015): 259–278.

Hellman, G. & Cook, R. T. (eds.) (2018) *Hilary Putnam on Logic and Mathematics*, Cham: Springer.

Hickey, L. P. (2009) *Hilary Putnam*, New York: Continuum.

Hildebrand, D. L. (2000) "Putnam, Pragmatism, and Dewey," *Transactions of the Charles S. Peirce Society* 36 (1): 109–132.

Hill, C. S. (ed.) (1992) "The Philosophy of Hilary Putnam," *Philosophical Topics* 20: 1–408.

Hodesdon, K. (2018) "The Metaphysics of the Model-Theoretic Arguments," in Hellman & Cook (2018): 75–91.

Hodges, A. (2019) "Alan Turing," in *The Stanford Encyclopedia of Philosophy*, Edward N. Zalta (ed.), https://plato.stanford.edu/entries/turing/.

Horgan, T. (ed.) (1991) "Special Issue on Putnam's Philosophy," *Erkenntnis* 34 (3): 269–424.

Horwich, P. (1998) *Truth*, Oxford: Clarendon Press; 1st edition Oxford: Basil Blackwell, 1990.

Horwich, P. (2016) "Wittgenstein on Truth," *Argumenta* 2 (1): 95–105.

Hovis, C. R. & Kragh, H. (1993) "P.A.M. Dirac and the Beauty of Physics," *Scientific American* 268: 104–109.

Hoyningen-Huene, P. (2011) "Reconsidering the Miracle Argument on the Supposition of Transient Underdetermination," *Synthese* 180: 173–187.

Hume, D. (1739–40) *A Treatise on Human Nature*, L. A. Selby-Bigge (ed.), Oxford: Clarendon Press, 1960.

Jackman, H. (2021) "Putnam, James, and 'Absolute Truth'," *European Journal of Pragmatism and American Philosophy* 13 (2). https://journals.openedition.org/ejpap/2509; in *Symposium. In Dialogue with Putnam: Pragmatism, Realism, and Normativity*, G. Marchetti (ed.).

Jacob, A. (eds.) (1989) *L'Encyclopédie philosophique universelle, Volume 1, L'univers philosophique*, Paris: PUF.

James, W. (1907) *Pragmatism*, Cambridge, MA: Harvard University Press, 1975.

Kant, I. (1998) *Critique of Pure Reason*, Cambridge: Cambridge University Press.

Kirchin, S. (ed.) (2013) *Thick Concepts*, Oxford: Oxford University Press.

Knowles, J. (2013) "Non-Reductive Naturalism and the Vocabulary of Agency," *Contemporary Pragmatism* 10 (2): 155–172.

Knowles, J. (2023) *Representation, Experience, and Metaphysics: Toward an Integrated Anti-Representationalist Philosophy*, Cham: Springer.

Kripke, S. (1972) "Naming and Necessity," in Harman & Davidson (1972): 253–355, 762–769; revised edition Oxford: Basil Blackwell, 1980.

Laugier, S. (2020) "Necrology of Ontology: Putnam, Ethics, Realism," *The Monist* 103, 391–403.

Lee, W.-Y. (2014) "Should the No-Miracle Argument Add to Scientific Evidence?," *Philosophia*, DOI 10.1007/s11406-014-9524-z.

Leng, M. (ed.) (2018) "Updating Indispensabilities: Putnam in Memoriam," *Theoria* 33 (2): 161–263.

Lepore, E. (ed.) (1986) *Truth and Interpretation: Perspectives on the Philosophy of Donald Davidson*, Oxford: Blackwell.

Lewis, D. (1984) "Putnam's Paradox," *Australasian Journal of Philosophy* 62: 221–236.

Linnebo, Ø. (2018) "Putnam on Mathematics as Modal Logic," in Hellman & Cook (2018): 249–267.

Lipton, P. (2001) "Quests of a Realist," *Metascience* 10 (3): 347–353.

Lynch, M. P. (2009) *Truth as One and Many*, Oxford: Oxford University Press.

Lynch, M. P., Wyatt, J., Kim, J. & Kellen, N. (eds.) (2021) *The Nature of Truth: Classic and Contemporary Perspectives*, 2nd edition, Cambridge, MA: MIT Press.

Macarthur, D. (2004) "Putnam's Natural Realism and the Question of a Perceptual Interface," *Philosophical Explorations* 7 (2): 159–173.

Macarthur, D. (2017) "Introduction," in Putnam & Putnam (2017): 1–9.

Macarthur, D. (2018) "The Many Faces of Objectivity," *Análisis* 5 (1): 91–109.

Macarthur, D. (2019) "Pragmatic Naturalism the Authority of Reason, the Agrippan Trilemma and the Significance of Philosophising *in medias res*," in Giladi (2019): 271–289.

Macarthur, D. (2020) "Exploding the Realism-Antirealism Debate: Putnam contra Putnam," *The Monist* 103: 370–380.

Mackie, J. (1974) *The Cement of the Universe: A Study of Causation*, Oxford: Clarendon Press.

Maddalena, G. (2018) "Review of *Pragmatism As a Way of Life* by Hilary Putnam and Ruth Anna Putnam", *William James Studies* 14: 177–18.

Maddy, P. (2007) *Second Philosophy: A Naturalistic Method*, Oxford: Oxford University Press.

Maitra, K. (2003) *On Putnam*, Belmont: Thomson Wadsworth.

Marchetti, G. & Marchetti, S. (eds.) (2017) *Facts and Values: The Ethics and Metaphysics of Normativity*, New York: Routledge.

Marras, A. (2001) "On Putnam's Critique of Metaphysical Realism: Mind-Body Identity and Supervenience," *Synthese* 126 (3): 407–426.

Maudlin, T. (2022) "The Labyrinth of Quantum Logic," in Conant & Chakraborty (2022): 183–205.

McDowell, J. (1994) *Mind and World*, Cambridge, MA: Harvard University Press.

McDowell, J. (1999) "Moderne Auffassungen von Wissenschaft und die Philosophie des Geistes," *Neue Rundschau* 110 (3): 48–69; repr. as "Naturalism in the Philosophy of Mind," in De Caro & Macarthur (2004): 91–105.

Mendelson, E. (2015) *Introduction to Mathematical Logic*, Boca Raton: Taylor & Francis.

Menke, C. (2014) "Does the Miracle Argument Embody a Base Rate Fallacy?," *Studies in History and Philosophy of Science* 45: 103–108.

Miguens, S. (2020) "The Human Face of Naturalism: Putnam and Diamond on Religious Belief and the 'Gulfs between Us'," *The Monist* 103: 404–414.

Moran, D. (2000) "Hilary Putnam and Immanuel Kant: Two 'Internal Realists'?," *Synthese* 123: 65–104.

Nagel, E., Suppes, P. & Tarski, A. (eds.) (1962) *Logic, Methodology and Philosophy of Science*, Stanford: Stanford University Press.

Narboux, J.-P. (2020) "Conceptual Truth, Necessity, and Negation," *The Monist* 103: 468–480.

Neustein, J. (2000) *Five Ash Cities*, Chicago: Academy Chicago Publishers.

Norris, C. (2002), *Hilary Putnam: Realism, Reason and the Use of Uncertainty*, Manchester: Manchester University Press.

Nussbaum, M. (2022) "Putnam's Aristotle," in Conant & Chakraborty (2022): 227–248.

Panza, M. & Sereni, A. (2013) *Plato's Problem: An Introduction to Mathematical Platonism*, Basingstoke: Palgrave Macmillan.

Parsons, C. (2015) "Putnam on Realism and 'Empiricism' in Mathematics," in Auxier, Anderson & Hahn (2015): 113–133.

Pearce, G. & Maynard, P. (eds.) (1973) *Conceptual Change*, Dordrecht: Reidel.

Pedersen, N. J. L. L. & Wright, C. (eds.) (2013) *Truth and Pluralism: Current Debates*, Oxford: Oxford University Press.

Pedersen, N. J. L. L. & Wright, C. (2016) "Pluralist Theories of Truth," in *The Stanford Encyclopedia of Philosophy*, E. N. Zalta (ed.), https://plato.stanford.edu/archives/spr2016/entries/truth-pluralist/.

Pessin, A. & Goldberg, S. (eds.) (1996) *The Twin-Earth Chronicles: Twenty Years of Reflection on Hilary Putnam's "The Meaning of 'Meaning'"*, Armonk: Sharpe; repr. Abingdon: Routledge, 2015.

Piattelli Palmarini, M. (ed.) (1984) *Livelli di realtà*, Milano: Feltrinelli.

Proctor, J. D. (ed.) (2005) *Science, Religion, and the Human Experience*, Oxford: Oxford University Press.

Psillos, S. (1999) *Scientific Realism: How Science Tracks Truth*, London: Routledge.

Psillos, S. (2001) "Quests of a Realist – Author's Response," *Metascience* 10 (3): 366–371.

Putnam, H. (1960a) "Do True Assertions Correspond to Reality?," in Putnam (1975b): 70–84.

Putnam, H. (1960b) "Minds and Machines," in Putnam (1975b): 362–385.

Putnam, H. (1962a) "It Ain't Necessarily So," in Putnam (1975a): 237–249.

Putnam, H. (1962b) "The Analytic and the Synthetic," in Putnam (1975b): 33–69.

Putnam, H. (1962c) "What Theories Are Not," in Nagel, Suppes & Tarski (1962): 240–252; repr. in Putnam (1975a): 215–227.

Putnam, H. (1963) "Brains and Behavior," in Butler (1963): 211–235; repr. in Putnam (1975b): 325–341.

Putnam, H. (1965) "A Philosopher Looks at Quantum Mechanics," in Colodny (1965): 75–101; repr. in Putnam (1975a): 130–158.

Putnam, H. (1967a) "Mathematics Without Foundations," *The Journal of Philosophy* 64: 5–22; in Putnam (1975a): 43–59.

Putnam, H. (1967b) "Psychological Predicates," in Capitan & Merrill (1967): 37–48; repr. as "The Nature of Mental States," in Putnam (1975b): 429–440.

Putnam, H. (1967c) "The Mental Life of Some Machines," in Castañeda (1967): 177–200; repr. in Putnam (1975b): 408–428.

Putnam, H. (1969a) "Logical Positivism and the Philosophy of Mind," in Achinstein & Barker (1969): 211–225; repr. in Putnam (1975b): 441–451.

Putnam, H. (1969b) "On Properties," in Rescher (1969): 235–254; repr. in Putnam (1975a): 305–322.

Putnam, H. (1970) "Is Semantics Possible?," *Metaphilosophy* 1: 187–201; repr. in Putnam (1975b): 139–152.

Putnam, H. (1971) *Philosophy of Logic*, New York: Harper & Row; repr. in Putnam (1975a), 2nd edition: 323–357.

Putnam, H. (1973) "Explanation and Reference," in Pearce & Maynard (1973): 199–221; repr. in Putnam (1975b): 196–214.

Putnam, H. (1974) "The 'Corroboration' of Theories," in Schilpp (1974): 221–240; repr. in Putnam (1975a): 250–269.

Putnam, H. (1974–75) "Reply to Lugg," *Cognition* 3: 295–298.

Putnam, H. (1975a) *Mathematics, Matter and Method*, *Philosophical Papers*, Volume 1, Cambridge: Cambridge University Press.

Putnam, H. (1975b) *Mind, Language and Reality*, *Philosophical Papers*, Volume 2, Cambridge: Cambridge University Press.

Putnam, H. (1975c) "Language and Reality," in Putnam (1975b): 272–290.

Putnam, H. (1975d) "Philosophy and Our Mental Life," in Putnam (1975b): 291–303.

Putnam, H. (1975e) "The Meaning of 'Meaning'," in Gunderson (1975): 131–193; repr. in Putnam (1975b): 215–271.

Putnam, H. (1975f) "What Is Mathematical Truth?," *Historia Mathematica* 2: 529–543; in Putnam (1975a): 60–78.

Putnam, H. (1975–1976) "What Is 'Realism'?," *Proceedings of the Aristotelian Society*, New Series, 76: 177–194.

Putnam, H. (1976a) "'Two Dogmas' Revisited," in Putnam (1983): 87–97.

Putnam, H. (1976b) "Literature, Science, and Reflection," *New Literary History* 7: 483–491; repr. in Putnam (1978a): 83–94.

Putnam, H. (1977) "Realism and Reason," *Proceedings and Addresses of the American Philosophical Association* 50 (6): 483–498; repr. in Putnam (1978a): 123–140.

Putnam, H. (1978a) *Meaning and the Moral Sciences*, London: Routledge & Kegan Paul.

Putnam, H. (1978b) "Equivalence," in Putnam (1983): 26–45.

Putnam, H. (1978c) "Reference and Understanding," in Putnam (1978a): 97–119.

Putnam, H. (1978d) "There Is At Least One *A Priori* Truth," in Putnam (1983): 98–114.

Putnam, H. (1980) "Models and Reality," *Journal of Symbolic Logic* 45: 464–482; repr. in Putnam (1983a): 1–25.

Putnam, H. (1981) *Reason, Truth and History*, Cambridge: Cambridge University Press.

Putnam, H. (1983a) *Realism and Reason, Philosophical Papers*, Volume 3, Cambridge: Cambridge University Press.

Putnam, H. (1983b) "On Truth," in Cauman, Levi, Parsons & Schwartz (1983): 35–56; repr. in Putnam (1994a): 315–329.

Putnam, H. (1983c) "Why There Isn't a Ready-Made World," in Putnam (1983a): 205–228.

Putnam, H. (1984a) "Is the Causal Structure of the Physical Itself Something Physical?," in French, Uehling & Wettstein (1984): 3–16; repr. in Putnam (1990a): 80–95.

Putnam, H. (1984b) "Realismo e relativismo concettuale: il problema del fatto e del valore," in Piattelli Palmarini (1984): 39–53.

Putnam, H. (1985) "After Empiricism," in Rajchman & West (1985): 20–30; repr. in Putnam (1990a): 43–53.

Putnam, H. (1986) "Meaning Holism," in Hahn & Schilpp (1986): 405–426; repr. in Putnam (1990a): 278–302.

Putnam, H. (1987a) *The Many Faces of Realism*, La Salle: Open Court.

Putnam, H. (1987b) "Truth and Convention: On Davidson's Refutation of Conceptual Relativism," *Dialectica* 41 (1–2): 69–77; repr. as "Truth and Convention," in Putnam (1990a): 96–104.

Putnam, H. (1988) *Representation and Reality*, Cambridge, MA: MIT Press.

Putnam, H. (1989a) "An Interview with Professor Hilary Putnam: The Vision and Arguments of a Famous Harvard Philosopher," *Cogito* 3 (2): 85–91; repr. in Pyle (1999): 44–54.

Putnam, H. (1989b) "Model Theory and the 'Factuality' of Semantics," in George (1989): 213–232.

Putnam, H. (1989c) "Pourquoi des philosophes?," in Jacob (1989): 765–771; repr. in Putnam (1990): 105–119.

Putnam, H. (1990a) *Realism with a Human Face*, J. Conant (ed.), Cambridge, MA: Harvard University Press.

Putnam, H. (1990b) "Is Water Necessarily H_2O?," in Putnam (1990a): 54–79.

Putnam, H. (1990c) "Objectivity and the Science Ethics Distinction," in Putnam (1990a): 163–178.

Putnam, H. (1991) "Does the Disquotational Theory Really Solve All Philosophical Problems?," *Metaphilosophy* 22: 1–13; repr. in Putnam (1994a): 264–278.

Putnam, H. (1992a) *Il pragmatismo: Una questione aperta*, Roma-Bari: Laterza; Engl. edition Oxford: Blackwell, 1995.

Putnam, H. (1992b) *Renewing Philosophy*, Cambridge, MA: Harvard University Press.

Putnam, H. (1992c) "Reply to Akeel Bilgrami," in Hill (1992): 385–391.

Putnam, H. (1992d) "Reply to Alan Sidelle," in Hill (1992): 391–393.

Putnam, H. (1992e) "Reply to Burton Dreben," in Hill (1992): 393–399.

Putnam, H. (1992f) "Reply to David Anderson," in Hill (1992): 361–369.

Putnam, H. (1992g) "Reply to Gary Ebbs," in Hill (1992): 347–358.

Putnam, H. (1992h) "Reply to Gerald Massey," in Hill (1992): 399–402.

Putnam, H. (1992i) "Reply to James Conant," in Hill (1992): 374–377.

Putnam, H. (1992j) "Reply to John McDowell," in Hill (1992): 358–361.

Putnam, H. (1992k) "Reply to Noam Chomsky," in Hill (1992): 379–385.

Putnam, H. (1992l) "Reply to Richard Healey," in Hill (1992): 377–379.

Putnam, H. (1992m) "Reply to Richard Miller," in Hill (1992): 369–374.

Putnam, H. (1993) "Realism without Absolutes," *International Journal of Philosophical Studies* 1 (2): 179–192; repr. in Putnam (1994a): 279–294.

Putnam, H. (1994a) *Words and Life*, J. Conant (ed.), Cambridge, MA: Harvard University Press.

Putnam, H. (1994b) "Pragmatism and Moral Objectivity," in Putnam (1994a): 151–181.

Putnam, H. (1994c) "Putnam, Hilary," in Guttenplan (1994): 507–513.

Putnam, H. (1994d) "Rethinking Mathematical Necessity," in Putnam (1994a): 245–263.

Putnam, H. (1994e) "Sense, Nonsense, and the Senses: An Inquiry into the Powers of the Human Mind," *Journal of Philosophy* 91: 445–517; repr. in Putnam (1999): 3–70.

Putnam, H. (1994f) "Simon Blackburn on Internal Realism," in Clark & Hale (1994): 242–254.

Putnam, H. (1994g) "The Question of Realism," in Putnam (1994a): 295–312.

Putnam, H. (1994h) "Between the New Left and Judaism: Hilary Putnam," in Borradori (1994): 55–69.

Putnam, H. (1994i) "Crispin Wright on the Brain-in-a-Vat Argument," in Clark & Hale (1994): 283–288.

Putnam, H. (1995) "Pragmatism," *Proceedings of the Aristotelian Society* 95: 291–306.

Putnam, H. (1996) "Introduction," in Pessin & Goldberg (1996): xv–xxii.

Putnam, H. (1997) "Interview with Hilary Putnam," G. Marchetti (ed.), *Cogito* 11 (3): 149–157.

Putnam, H. (1999) *The Threefold Cord: Mind, Body, and World*, New York: Columbia University Press.

Putnam, H. (2000a) "Richard Rorty on Reality and Justification," in Brandom (2000): 81–87.

Putnam, H. (2000b) "Thoughts about Domestic Tranquility / Bne Brak," in Neustein (2000): 100–108.

Putnam, H. (2001) "Reply to Jennifer Case," *Revue internationale de philosophie* 218: 431–438.

Putnam, H. (2002a) *The Collapse of the Fact/Value Dichotomy and Other Essays*, Cambridge, MA: Harvard University Press.

Putnam, H. (2002b) "Reply to Ruth Anna Putnam," in Conant & Zeglen (2002): 12–13; repr. in Putnam & Putnam (2017): 18–20.

Putnam, H. (2003) "For Ethics and Economics Without the Dichotomies," *Review of Political Economy* 15: 305–412; repr. in Putnam & Walsh (2012): 111–129.

Putnam, H. (2004a) *Ethics Without Ontology*, Cambridge, MA: Harvard University Press.

Putnam, H. (2004b) "The Content and Appeal of 'Naturalism'," in De Caro & Macarthur (2004): 59–70.

Putnam, H. (2005a) "The Depths and Shallows of Experience," in Proctor (2005): 71–86; repr. in Putnam (2012a): 567–583.

Putnam, H. (2005b) "Jewish Ethics," in Schweiker (2005): 159–165.

Putnam, H. (2008a) *Jewish Philosophy as a Guide to Life: Rosenzweig, Buber, Levinas, Wittgenstein*, Bloomington and Indianapolis: Indiana University Press.

Putnam, H. (2008b) "Reply to David Macarthur," *European Journal of Analytic Philosophy* 4 (2): 47–49.

Putnam, H. (2008c) "Reply to Massimo Dell'Utri," *European Journal of Analytic Philosophy* 4 (2): 87–91.

Putnam, H. (2008d) "Reply to Mauro Dorato," *European Journal of Analytic Philosophy* 4 (2): 71–73.

Putnam, H. (2008e) "Reply to Stephen White," *European Journal of Analytic Philosophy* 4 (2): 29–32.

Putnam, H. (2008f) "12 Philosophers–and Their Influence on Me," *Proceedings and Addresses of the American Philosophical Association* 82 (2): 101–115; repr. in Putnam (2022): 327–344.

Putnam, H. (2010) "Science and Philosophy," in De Caro & Macarthur (2010): 89–99.

Putnam, H. (2011) "The Story of Pragmatism," *Comprendre* 13 (1): 37–48.

Putnam, H. (2012a) *Philosophy in an Age of Science: Physics, Mathematics, and Skepticism*, M. De Caro & D. Macarthur (eds.), Cambridge, MA: Harvard University Press.

Putnam, H. (2012b) "Corresponding with Reality," in Putnam (2012a): 72–90.

Putnam, H. (2012c) "From Quantum Mechanics to Ethics and Back Again," in Putnam (2012a): 51–71 (quotations are from this book); and in Baghramian (2013): 19–36.

Putnam, H. (2012d) "How to Be a Sophisticated 'Naïve Realist'," in Putnam (2012a): 624–639.

Putnam, H. (2012e) "Indispensability Arguments in the Philosophy of Mathematics," in Putnam (2012a): 181–201.

Putnam, H. (2012f) "On Not Writing Off Scientific Realism," in Putnam (2012a): 91–108.

Putnam, H. (2012g) "Set Theory: Realism, Replacement, and Modality," in Putnam (2012a): 217–234.

Putnam, H. (2012h) "Wittgenstein: A Reappraisal," in Putnam (2012a): 482–492.

Putnam, H. (2013a) "Comments on Richard Boyd," in Baghramian (2013): 95–100.

Putnam, H. (2013b) "The Development of Externalist Semantics," *Theoria* 79 (3): 192–203; repr. in Putnam (2016a): 199–212.

Putnam, H. (2015a) "Against Perceptual Conceptualism," with Hilla Jacobson, *International Journal of Philosophical Studies* 24 (1): 1–25.

Putnam, H. (2015b) "Intellectual Autobiography," in Auxier, Anderson & Hahn (2015): 3–110.

Putnam, H. (2015c) "Naturalism, Realism, and Normativity," *Journal of the American Philosophical Association* 1 (2): 312–328; repr. in Putnam (2016a): 21–43.

Putnam, H. (2015d) "Reply to Frederick Stoutland," in Auxier, Anderson & Hahn (2015): 547–563.

Putnam, H. (2015e) "Reply to Geoffrey Hellman," in Auxier, Anderson & Hahn (2015): 279–282.

Putnam, H. (2015f) "Reply to Larry A. Hickman," in Auxier, Anderson & Hahn (2015): 788–800.

Putnam, H. (2015g) "Reply to Steven J. Wagner," in Auxier, Anderson & Hahn (2015): 240–258.

Putnam, H. (2015h) "Reply to Pierre Hadot," in Auxier, Anderson & Hahn (2015): 677–681.

Putnam, H. (2015i) "Reply to John Haldane," in Auxier, Anderson & Hahn (2015): 702–706.

Putnam, H. (2015l) "Reply to Ruth Anna Putnam," in Auxier, Anderson & Hahn (2015): 725–733.

Putnam, H. (2015m) "Reply to Cornel West," in Auxier, Anderson & Hahn (2015): 768–774.

Putnam, H. (2015n) "Reply to Richard Rorty," in Auxier, Anderson & Hahn (2015): 882–889.

Putnam, H. (2015o) "Interview with Hilary Putnam," M. Bella & A. Boncompagni (eds.), *European Journal of Pragmatism and American Philosophy* VII (1): 1–9.

Putnam, H. (2016a) *Naturalism, Realism, and Normativity*, M. De Caro (ed.), Cambridge, MA: Harvard University Press.

Putnam, H. (2016b) "Reading Rosenzweig's Little Book," *Argumenta* 1 (2): 161–168.

Putnam, H. (2016c) "Realism," *Philosophy and Social Criticism*, 42 (2): 117–131.

Putnam, H. (2022) *Philosophy as Dialogue*, M. De Caro & D. Macarthur (eds.), Cambridge, MA: Harvard University Press.

Putnam, H. & Bouveresse, J. (2020) "A Conversation," *The Monist* 103: 481–492.

Putnam, H. & Davis, M. (1958) "Reductions of Hilbert's Tenth Problem," *Journal of Symbolic Logic* 23 (2): 183–187.

Putnam, H., Davis, M. & Robinson, J. (1961) "The Decision Problem of Exponential Diophantine Equations," *Annals of Mathematics* 74 (3): 425–436.

Putnam, H. & Putnam, R. A. (2017) *Pragmatism as a Way of Life: The Lasting Legacy of William James and John Dewey*, D. Macarthur (ed.), Cambridge, MA: Harvard University Press.

Putnam, H. & Walsh, V. (2007) "Facts, Theories, Values and Destitution in the Works of Sir Partha Dasgupta," *Review of Political Economy* 19: 181–202; repr. in Putnam & Walsh (2012): 150–171.

Putnam, H. & Walsh, V. (eds.) (2012) *The End of Value-Free Economics*, New York: Routledge.

Putnam, R. A. (1985) "Creating Facts and Values," *Philosophy* 60: 187–204; repr. in Putnam & Putnam (2017): 385–404.

Putnam, R. A. (1987) "Weaving Seamless Webs," *Philosophy* 62: 207–220; repr. in Putnam & Putnam (2017): 71–86.

Putnam, R. A. (1998a) "Perceiving Facts and Values," *Philosophy* 73: 5–19; repr. in: Putnam & Putnam (2017): 405–420.

Putnam, R. A. (1998b) "Perception: From Moore to Austin," in Biletzki & Matar (1998): 182–196.

Putnam, R. A. (2002) "Taking Pragmatism Seriously," in Conant & Zeglen (2002): 7–13; repr. in Putnam & Putnam (2017): 13–18.

Putnam, R. A. (2013) "Hilary Putnam's Moral Philosophy," in Baghramian (2013): 240–256.

Putnam, R. A. (2015) "Hilary Putnam's Jewish Philosophy," in Auxier, Anderson & Hahn (2015): 707–724.

Putnam, R. A. (2017) "Reflections Concerning Moral Objectivity," in Marchetti & Marchetti (2017): 105–118.

Pyle, A. (ed.) (1999) *Key Philosophers in Conversation: The* Cogito *Interviews*, London: Routledge.

Quine, W. V. O. (1948) "On What There Is," in Quine (1953): 1–19.

Quine, W. V. O. (1951) "Two Dogmas of Empiricism," in Quine (1953): 20–46.

Quine, W. V. O. (1953) *From a Logical Point of View*, Cambridge, MA: Harvard University Press.

Quine, W. V. O. (1960) *Word and Object*, Cambridge, MA: MIT Press.

Quine, W. V. O. (1963) "Carnap and Logical Truth," in Schilpp (1963): 385–405; repr. in Quine (1966): 100–125.

Quine, W. V. O. (1966) *The Ways of Paradox and Other Essays*, New York: Random House.

Quine, W. V. O. (1968) "Ontological Relativity," *The Journal of Philosophy* 65: 185–212; repr. in Quine (1969): 26–68.

Quine, W. V. O. (1969) *Ontological Relativity and Other Essays*, New York: Columbia University Press.

Quine, W. V. O. (1981) *Theories and Things*, Cambridge, MA: Harvard University Press.

Quine, W. V. O. (1992) *Pursuit of Truth*, 2nd ed., Cambridge, MA: Harvard University Press.

Rajchman, J. & West, C. (eds.) (1985) *Post-Analytic Philosophy*, New York: Columbia University Press.

Reichenbach, H. (1938) *Experience and Prediction: An Analysis of the Foundations and the Structure of Knowledge*, Chicago: The University of Chicago Press.

Reichenbach, H. (1949) "The Philosophical Significance of the Theory of Relativity," in Schilpp (1949): 289–311.

Rescher, N. (ed.) (1969) *Essays in Honor of Carl G. Hempel*, Dordrecht: Reidel.

Richardson, A. (2006) "Rational Reconstruction," in Sarkar & Pfeifer (2006): 681–685.

Rivas Monroy, M. U., Cancela Silva, C. & Martínez Vidal, C. (eds.) (2008) "Following Putnam's Trail: On Realism and Other Issues," *Poznan Studies in the Philosophy of the Sciences and the Humanities* 95: 3–310.

Rorty, R. (1979) *Philosophy and the Mirror of Nature*, Princeton: Princeton University Press.

Rorty, R. (1985) "Solidarity or Objectivity?," in Rajchman & West (1985): 3–19; repr. in Rorty (1991): 21–34.

Rorty, R. (1986) "Pragmatism, Davidson and Truth," in Lepore (1986): 333–355; repr. in Rorty (1991): 126–150.

Rorty, R. (1989) *Contingency, Irony, and Solidarity*, Cambridge: Cambridge University Press.

Rorty, R. (1991) *Objectivity, Relativism, and Truth*, *Philosophical Papers*, Volume 1, Cambridge: Cambridge University Press.

Rorty, R. (1993) "Putnam and the Relativist Menace," in *Journal of Philosophy* 90: 443–461; repr. in Rorty (1998a): 43–62.

Rorty, R. (1994) "John Searle on Realism and Relativism," *Academe* 80: 52–63 (as "Does Academic Freedom Have Philosophical Presuppositions?"); repr. in Rorty (1998): 63–83.

Rorty, R. (1995) "Is Truth a Goal of Inquiry? Donald Davidson versus Crispin Wright," *Philosophical Quarterly* 45: 281–300; repr. in Rorty (1998a): 19–42.

Rorty, R. (1998a) *Truth and Progress*, *Philosophical Papers*, Volume 3, Cambridge: Cambridge University Press.

Rorty, R. (1998b) "The Very Idea of Human Answerability to the World: John McDowell's Version of Empiricism," in Rorty (1998a): 138–152.

Rorty, R. (1999) *Philosophy and Social Hope*, New York: Penguin.

Rorty, R. (2000) "Universality and Truth," in Brandom (2000): 1–30.

Rorty, R. (2015) "Putnam, Pragmatism, and Parmenides," in Auxier, Anderson & Hahn (2015): 863–881.

Rorty, R., Putnam, H., Conant, J. & Helfrich, G. (2004) "What Is Pragmatism?," *Think* 3: 71–88.

Rudner, J. (1983) *A Stroll with William James*, New York: Harper & Row.

Ryle, G. (1968) "The Thinking of Thoughts: What Is 'Le Penseur' Doing?," *University Lectures* 18; repr. in Ryle (1971): 494–510.

Ryle, G. (1971) *Collected Papers*, Volume 2, Abingdon: Routledge, 2009.

Sarkar, S. & Pfeifer, J. (eds.) (2006) *The Philosophy of Science: An Encyclopedia*, Abingdon: Routledge.

Schilpp, P. A. (ed.) (1949) *Albert Einstein: Philosopher-Scientist*, New York: Tudor.

Schilpp, P. A. (ed.) (1963) *The Philosophy of Rudolf Carnap*, La Salle: Open Court.

Schilpp, P. A. (ed.) (1974) *The Philosophy of Karl Popper*, La Salle: Open Court.

Schwartz, R. (2021) "Putnam and the Pragmatists," *European Journal of Pragmatism and American Philosophy* 13 (2). http://journals.openedition.org/ejpap/2502; in *Symposium. In Dialogue with Putnam: Pragmatism, Realism, and Normativity*, G. Marchetti (ed.).

Schwartz, S. P. (1978) "Putnam on Artifacts," *Philosophical Review* 87 (4): 566–74; repr. in Pessin & Goldberg (1996): 81–88.

Schweiker, W. (ed.) (2005) *The Blackwell Companion to Religious Ethics*, Oxford: Blackwell.

Searle, J. R. (1995) *The Construction of Social Reality*, New York: The Free Press.

Sellars, W. (1962) "Philosophy and the Scientific Image of Man," in Colodny (1962): 35–78.

Shagrir, O. (2005) "The Rise and Fall of Computational Functionalism," in Ben-Menahem (2005): 220–250.

Shook, J. R. & Margolis, J. (eds.) (2006) *A Companion to Pragmatism*, Malden: Blackwell.

Skolem, T. A. (1967) "Some Remarks on Axiomatized Set Theory," in van Heijenhoort (1967): 291–301.

Soysal, Z. (2020) "Descriptivism about the Reference of Set-Theoretic Expressions: Revisiting Putnam's Model-Theoretic Arguments," *The Monist*: 103, 442–454.

Stegmüller, W. (1987) *Hauptströmungen der Gegenwartsphilosophie*, Band II, Stuttgart: Kröner.

Stoutland, F. (2002) "Putnam on Truth," in Gustafsson & Hertzberg (2002): 147–176.

Thorpe, J. R. (2018a) "Closure Scepticism and the Vat Argument," *Mind* 127 (507): 667–690.

Thorpe, J. R. (2018b) "Review of Tim Button's *The Limits of Realism*," *Argumenta* 3 (2): 389–393.

Thorpe, J. R. (2019) "Semantic Self-Knowledge and the Vat Argument," *Philosophical Studies* 176: 2289–2306.

Thorpe, J. R. & Wright, C. (2022) "Putnam's Proof Revisited," in Conant & Chakraborty (2022): 63–88.

Tiercelin, C. (2002) *Hilary Putnam: L'héritage pragmatiste*, Paris: PUF.

Timmons, M. (1991) "Putnam's Moral Objectivism," *Erkenntnis* 34 (3): 371–399.

Tomberlin, J. E. (ed.) (1997) *Mind, Causation, and World*, Malden: Blackwell.

Travis, C. (2005) "The Face of Perception," in Ben-Menahem (2005): 53–82.

Travis, C. (2020) "The Objects of Rational Thought (a Sketch)," *The Monist* 103: 455–467.

Turing, A. (1936) "On Computable Numbers, With an Application to the Entscheidungsproblem," *Proceedings of the London Mathematical Society Ser* 2 (42): 230–265.

Tymoczko, T. (1989) "In Defense of Putnam's Brains," *Philosophical Studies* 57: 281–297.

van Fraassen, B. (1997) "Putnam's Paradox: Metaphysical Realism Revamped and Evaded," in Tomberlin (1997): 17–42.

van Heijenhoort, J. (ed.) (1967) *From Frege to Gödel*, Cambridge, MA: Harvard University Press.

Wagner, H. (2020) "The Significance of the Division of Linguistic Labor," *The Monist* 103: 381–390.

Wagner, S. J. (2015) "Modal and Objectual," in Auxier, Anderson & Hahn (2015): 221–239.

Walsh, V. (1987) "Philosophy and Economics," in Eatwell, Milgate & Newman (1987): 861–869.

West, C. (2015) "Hilary Putnam and the Third Enlightenment," in Auxier, Anderson & Hahn (2015): 757–767.

Williams, B. (1985) *Ethics and the Limits of Philosophy*, Cambridge, MA: Harvard University Press.

Williams, M. (1993) "Realism and Scepticism," in Haldane & Wright (1993): 193–214.

Wittgenstein, L. (1953) *Philosophical Investigations*, Oxford: Basil Blackwell.

Wittgenstein, L. (1966) *Lectures and Conversations on Aesthetics, Psychology, and Religious Belief*, C. Barrett (ed.), Berkeley: University of California Press.

Wittgenstein, L. (1969) *On Certainty*, Oxford: Basil Blackwell.

Wittgenstein, L. (1976) "Cause and Effect: Intuitive Awareness," *Philosophia* 6 (3–4): 409–425.

Wright, C. (1994) "On Putnam's Proof that We Are Not Brains in a Vat", in Clark & Hale (1994): 216–241.

Wright, C. (2000) "Truth as Sort of Epistemic: Putnam's Peregrinations," *The Journal of Philosophy* 97: 335–364.

Wright, C. (2001) "Minimalism, Deflationism, Pragmatism and Pluralism," in Lynch (2001): 751–787.

Zahavi, D. (2004) "Natural Realism, Anti-Reductionism, and Intentionality: The 'Phenomenology' of Hilary Putnam," in Carr & Cheung (2004): 235–251.

Index

Adorno, Theodor, 95
Alai, Mario, 7, 92n2
Albritton, Rogers, 88
alethic deflationism, 148, 158nn18–19, 159n20, 196–97, 204
analyticity, 9, 10, 12–13, 23nn1–2, 72, 79n20, 209
anti-reductionism, 60
anti-scientism, 60
anti-utopianism, 60
a priori, 5, 9, 13–17, 19–25, 27, 31, 56, 72–74, 153–54, 168, 190, 211; contextually, 21, 25; relatively, 27. *See also* contextually a priori; synthetic, 20, 27, 154, 211. *See also* contextually a priori
Archimedes, 121n2
Aristotle, 55, 64, 76n4, 77n5, 158n15, 163, 207n8
Artuso, Paolo, 7
Audi, Robert, 7
Austin, John L., 116, 121n7, 128–29, 131
Ayer, Alfred J., 207n5

Bacchini, Fabio, 7
Bacon, Francis, 26
Baghramian, Maria, xv, 7, 93, 208n21
Bagnoli, Carla, 7

Bangu, Sorin, 50n18
Bays, Timothy, 122n15
behaviorism, 51, 52, 62n1
Benacerraf, Paul, 44, 49n15
Ben-Menahem, Yemima, 61
Berkeley, George, 83, 87
Berlin, Isaiah, 188n20
Bernecker, Sven, 152
Bernstein, Richard J., 186
Berti, Enrico, 158n15
Biondi, Deanna, 7
Blackburn, Simon, 125
Bohr, Niels, 90–91
Bolyai, János, 15
Borradori, Giovanna, 206n1
Bourbaki, Nicolas, 42
Bouveresse, Jacques, 8n6
Boyd, Richard, 89, 92n2, 115, 122n16
Braddon-Mitchell, David, 7n1
brains in a vat, 28, 95–97, 111, 117–20, 124nn24–25, 150, 152–55, 159nn24–26, 171, 172
Brandom, Robert, 208n21
Brueckner, Anthony, 153
Buber, Martin, 177–81
Bueno, Otávio, 50n18
Burge, Tyler, 79n22
Burgess, John P., 49n16
Button, Tim, 124n24, 157n7, 159n24

233

Calcaterra, Rosa M., 7
Campanella, Tommaso, 210
Cantor, Georg, 104
Caputo, Stefano, 7, 188n18
Carnap, Rudolf, x, 1, 32–33, 46, 49n8, 50n17, 62nn1–6, 64, 83, 85, 92n1, 181, 204, 211
Case, Jennifer, 36–38, 208n18
causality, 115–16, 123n18
Cavell, Stanley, x, 161, 162, 179, 207nn6–8
certainty, 16, 182
Chakraborty, Sanjit, 7, 76n1, 187n4
Chomsky, Noam, ix
Churchman, C. West, 190
Clark, Kelly James, 7n1
Coliva, Annalisa, 159n25
common sense, 34, 133, 135, 141, 177, 202, 203, 208n19
Compagnon, Antoine, 187n11
Conant, James, 3, 192, 204, 207nn3–8, 208n20, 209
constraints: operational, 105–6, 108–10, 113, 121n8, 122n8, 129; theoretical, 105, 108–10, 113, 121n8
convergence of theories, 92n4, 199
cookie cutter, 31–32, 34, 107
Cormier, Harvey, 207, 208n21
correspondence, 105, 113, 115, 122n16, 156, 185, 202; truth as, 94–96, 103–4, 106–9, 111–13, 142–43, 145–47, 158n17, 195–96
Corvi, Roberta, 7

Dalton, John, 91
Dancy, Jonathan, 187n2
Darwin, Charles, 84, 197, 208n19
Davidson, Donald, 34, 126, 165
Davies, Richard, 7, 121n3
Davis, Martin, x–xii, 48n1
De Caro, Mario, 4, 7, 7n4, 48n4, 210
Dedekind, Richard, 46
deflationism about truth. *See* alethic deflationism
de Gaynesford, Maximilian, 8n6

Dell'Utri, Massimo, 7n4, 119
De Mol, Liesbeth, 53
Dennett, Daniel, 122n16
Descartes, René, 25, 95, 118, 120, 121n3, 128–29, 154
Devitt, Michael, 122n16
Dewey, John, 3, 61, 94–95, 116, 128, 162, 178–79, 190–91, 200, 206nn2–3, 207n6, 208n19
Diamond, Cora, 207n8
dichotomy: actually existing/mere projection, 116–17; analytic/synthetic, 4, 9, 10, 20, 23n1, 32, 49n8, 117, 170–71; fact/value, 166, 169; internal/external, 117; objective/subjective, 116–17; observational/theoretical, 88, 117, 170; true/assertible, 117
Diesendruck, Erna, x
Dirac, Paul, 123n17, 187n7
disquotation, 119, 148, 150, 158n18, 159n20, 196–97
do Carmo, Juliano, 7, 7n1
dogma of empiricism, 4, 9, 10, 13, 14; the third, 22
Douven, Igor, 122n15
Dummett, Michael, 94, 98, 108, 110–12, 126, 134, 136, 140, 157nn7–8
Dümont, Jürgen, 122n15

Ebbs, Gary, 81, 124n24, 135, 154
Egidi, Rosaria, 7
Einstein, Alfred, 15, 28, 173
Elgin, Catherine Z., 187n5
epistemic values, 171, 173, 174, 187n6
equivalence thesis, 119–20, 146, 148, 158n18, 159n20, 197
equivalent descriptions, 28, 30–31, 34–38, 40, 43–45, 47, 48nn5–6, 49n16, 97, 101
essence, 75, 95, 123n21
essentialism, 75, 123n21
ethics, function of, 163
Euclid, 15–16, 24n8
Euler, Leonhard, 26

externalism, xii, 62–64, 76n1, 78nn13–15, 81, 90, 93, 131

fact and convention, interpenetration of, 11, 31–34, 39, 49n8, 107, 144, 170–71
fact and value, entanglement of, 181, 191
fallacy of division, 107
fallibilism, 13, 20–23, 100, 191, 210
Feigl, Herbert, xi, 62n1
Fermat, Pierre de, 41–42
Fernández-Vega, José M. S., 50n18
Ferrari, Filippo, 7, 159n23
Feyerabend, Paul K., 81, 117
Feynman, Richard, 68
Fiorato, Pierfrancesco, 7
Fitch, Frederic B., 157n10
Floyd, Juliet, 7n4, 49n16, 64, 77nn6–8, 78n13, 79n18
Ford, Ford Madox, ix
Foucault, Michel, 198
Fraenkel, Abraham A. H., xi, 104
Frege, Gottlob, 42, 64, 126, 148, 158n19
French, Peter A., 7n1
Freud, Sigmund, 1, 207nn5–14
Friggieri, Joe, 7
functionalism, xi, xiii–xiv, 5, 51, 52, 54–56, 58–61, 62n4, 129–32

García-Carpintero, Manuel, 122n15
Gauss, Carl, 15
Geertz, Clifford, 187n8
Giametta, Sossio, 7
Giladi, Paul, 7n1
Gil Martín, Francisco J., 207n9
Gilson, Étienne-Henri, 206
Glymour, Clark, 122n16
Gödel, Kurt, 97, 122n14
God's eye perspective, 6, 95, 100, 110, 116, 122n12, 143, 184, 195, 203
Goldberg, Sanford C., 77n10, 123n22
Goodman, Nelson, 121n1
Goodman, Russell B., 207n6
Grice, Paul, 10–11, 17, 23n3

Habermas, Jürgen, 187n4, 207nn8–9
Hadot, Pierre, 163, 176–77, 181, 186n1
Haldane, John, 187n12, 188n13
Hale, Bob, 122n15
Hanson, Norwood R., 211
Hartz, Carolyn G., 208n18
Hawking, Stephen, 137
Hegel, Georg W. F., 95
Heidegger, Martin, 151
Helfrich, Gretchen, 207n3
Hellman, Geoffrey, 41, 42, 45, 49n15
Hempel, Carl G., 62n1
Hilbert, David, xi–xii
Hildebrand, David L., 207n7
Hitler, Adolf, 2
Hodesdon, Kate, 122n15
Hodges, Andrew, 53
Hodges-Gluck, Jana, 7
holism, 14, 202
Horwich, Paul, 148–49, 158n19
Hovis, R. Corby, 187n7
Hoyningen-Huene, Paul, 92n2
Hume, David, 76n4, 114, 116, 167–70; Hume's law, 167–68, 170
Husserl, Edmund, 116

ideal epistemic conditions, 99, 102, 127, 128
idealism, 18, 39, 82–83, 85–86, 88, 97, 99, 113, 121n5
idealistic fallacy, 86, 196
identity, theory of, 51, 52, 56–58, 62nn1–2
indexicality, 72–74
indispensability argument, 43, 48, 50n18, 192–94, 207n9
intentionality, 61, 94, 98, 115, 125, 128, 131, 198
interface, 128–29
intuition, 26, 70, 101–2, 133, 136–37, 156n2, 203, 208n19

Jackman, Henry, 207n3
Jacobs, Ruth Anna. *See* Putnam, Ruth Anna

Jacobson, Hilla, 156n5, 157n5
James, William, 3, 94, 116, 121n1, 128–
    29, 156n5, 178–79, 190–91, 196,
    199–200, 205, 206nn2–3, 207nn3–6

Kant, Immanuel, 20–21, 30, 33, 49n10,
    94, 113, 121nn4–5, 151, 163,
    188n15, 207n8
Kellen, Nathan, 159n22
Kierkegaard, Søren, 1, 177, 187n10,
    207n5
Kim, Junyeol, 159n22
Kirchin, Simon, 187n8
Knowles, Jonathan, 7, 153
Kragh, Helge, 187n7
Kreisel, Georg, x, 44
Kripke, Saul, 22, 68–69, 71–75,
    78nn15–16
Kuhn, Thomas S., 81, 211

Latour, Bruno, 198
Laugier, Sandra, 188n19
Lebenswelt, 116
Lee, Wang-Yen, 92n2
Leibniz, Gottfried W. von, 26, 33
Leng, Mary C., 50n18
Leśniewski, Stanislaw, 32–33, 39
Levinas, Emmanuel, 163–64, 177, 180
Lewis, Clarence I., 207n6
Lewis, David, 122n15–16
Linnebo, Øystein, 49n16
Lipton, Peter, 92n2
Lizzadri, Antonio, 7
Lobachevsky, Nikolai, 15
logical positivism, 1–5, 16, 27, 46,
    81–86, 88, 90, 108, 121n8, 123n19,
    136, 138, 169–70, 173, 190
Lorentz, Hendrik A., 29
Löwenheim, Leopold, 104–6, 111–12,
    117
Lynch, Michael P., 159n22

Macarthur, David, 4, 7, 7n4, 156n4,
    188n21, 190, 210
Mach, Ernst, 83

Mackie, John, 122n17, 207n11
Maddy, Penelope, 46–47, 207n10
Marchetti, Giancarlo, 7
Marchetti, Sarin, 7
Marconi, Diego, 7
Martínez-Vidal, Concha, 50n18
Marx, Karl H., 1, 207n5
materialism, 83, 113–16, 204
Maudlin, Tim, 8n6
McDowell, John, 4, 7n3, 62n8, 128–29,
    156nn3–5, 207n8
meaning, 5, 9, 17–18, 23n6, 35–37, 41,
    56–57, 61, 63–74, 76n2, 77nn9–11,
    78n12, 82–84, 90–91, 93, 126–27,
    131, 133, 147, 170; vector of, 71,
    79n18
Mendel, Gregor J., 91
Mendelson, Elliott, 122n14
Menke, Cornelis, 92n2
Mercator, Gerardus, 31
Miguens, Sofia, 187n9
Mill, John S., 43, 49n13, 64, 114
Millikan, Ruth, 131
mind, 51–55, 57, 59–62, 62n8, 64,
    76n4, 78n13, 118, 129–33, 146, 150,
    155, 192
modalism, 5, 28, 40–45, 47–48,
    49nn15–16
model-theoretic argument, 28, 103–6,
    109–12, 121n6, 122n15, 132
Moran, Dermot, 121n5
Moruzzi, Sebastiano, 7
multiple realizability, 55, 59, 61, 62n4

Nagel, Thomas, 95
Narboux, Jean-Philippe, 23n7
naturalism, 1–2, 4, 6, 7nn1–4, 55, 60,
    63, 93, 176–77, 181, 186n1, 187n12,
    188n13, 189, 193, 205–6, 207n4, 209
Navia, Ricardo, 7
necessity, 5, 9, 10, 13–17, 19–22, 23n5,
    25, 51, 72–76, 192; a posteriori,
    73–75; quasi-, 19–21, 23n7, 153–55,
    159n27, 211; relative. *See* quasi-
    necessity

Newton, Isaac, 26, 84, 123n19, 207n14
Nola, Robert, 7n1
nominalism, 27, 76n4
No-Miracle Argument, 88–89, 92n2
non-Euclidean geometries, 15, 17–19
normativity, 91, 103, 147–49, 159n23,
    173–75, 193
Nussbaum, Martha C., 55

object, 18, 37, 128, 144–45, 147
objectivity, 27, 44, 48, 76, 103, 107,
    143, 165–66, 179, 182–86, 187n12,
    188nn18–21, 193–97, 199
Ockham, William of, 122n13, 172
Onofri, Massimo, 7
ontology, 30, 33–34, 38–39, 45, 142,
    145, 188n19

Panza, Marco, 50n18
Parsons, Charles, 48n1
Passmore, John, 209–10
Paternoster, Alfredo, 7
Peano, Giuseppe, 27
Pedersen, Nikolaj J. L. L., 159n22
Peirce, Charles S., 97, 102, 123n20,
    137, 144, 156, 173, 190–91, 200,
    206nn2–3
Penco, Carlo, 7
perception, 128–32, 134, 150, 156nn3–
    5, 157n5, 188n15, 204
Perconti, Pietro, 7
Perissinotto, Luigi, 7, 159n27
Pesce, Isabella, 7
Pessin, Andrew, 77n10
Philo of Alexandria, 177
philosophy, 2, 6, 94, 98, 113, 156, 161–
    63, 176–77, 181, 186, 186n1, 189,
    192–93, 206, 206n1, 210–11
physicalism, 113–14, 122n16
Piatt, Donald, 190
Pirandello, Luigi, ix
Place, Ullin T., 62n1
Plato, 76n4, 116
Platonism, 27–28, 40, 42–45, 47–48,
    104, 192

Plebani, Matteo, 50n18
pluralism: alethic, 146–50; conceptual,
    31, 38
Popper, Karl R., 81
pragmatism, 3, 13, 94, 97–98, 100, 116,
    162, 173, 187n4, 190–96, 199–200,
    205, 206n2, 207nn4–6, 208nn19–21
principle of the benefit of the doubt,
    90–91
principle of verification, 82
Psillos, Stathis, 92n2
Putnam, Erika, x
Putnam, Joshua, xii
Putnam, Maxima. *See* Putnam, Polly
Putnam, Polly, xii
Putnam, Ruth Anna, xi–xii, xvi, 158n14,
    177, 179, 181–84, 187nn3–6,
    188n17, 200–201
Putnam, Samuel, ix
Putnam, Samuel jr., xii, 7

Quasi-empirical, 25–27
Quine, Willard V. O., x, 4–5, 9–14,
    21–22, 23nn1–2, 24n9, 34, 48, 48n5,
    49n9, 79n20, 81, 95, 100, 121n6,
    122n16, 148, 158nn15–18, 170–71,
    192–93, 199, 211

Rabelais, François, ix
Rainone, Antonio, 7
Ramsey, Frank P., 148
rationality, 13, 21, 23, 73, 81, 107, 133,
    168, 185, 201
realism, 7n5, 44, 48, 88, 91–94, 101,
    111–12, 120, 121n4, 125, 134,
    135, 141, 142, 150, 204, 208n19;
    commonsense, 205; epistemic, 112,
    142; internal, xiii, xiv, 96–103,
    107, 111–12, 116–17, 120, 121n5,
    125–31, 134–35, 142, 156n2;
    metaphysical, 28, 48n3, 93–99, 103,
    113, 116, 120, 122nn15–16, 128,
    130, 142–45, 159n24, 205; natural,
    xiv, 125, 127, 129–30, 142, 146,
    149, 157n7; non-epistemic, 111, 120,

142, 150, 159n24; pragmatic, 100;
scientific, xii, 88–89, 91–92, 98–99;
sophisticated metaphysical, 97, 107
reference, 5, 18, 63–64, 68, 75, 78nn15–
17, 88–92, 106, 113, 115–16, 118–
19, 121, 122n10, 126, 133, 148–50,
156, 159n20, 185, 202–3
Reichenbach, Hans, x, 1, 16, 46, 48n5,
92n1
relativism, 48n4, 101, 103, 107, 117,
188n20, 201–2, 208n18
relativity: conceptual, 28, 30–31, 33–34,
37–40, 43, 45, 48n4, 101, 103,
107, 142, 144–45; ontological, 100,
121n6, 145; theory of, 1, 15, 28–29
religion, function of, 181
Richardson, Alan, 46
Riemann, Bernhard, 15
rigid designator, 72
Robinson, Julia, xii
Rorty, Richard, 133, 187n4, 190,
194–205, 206n2, 207nn3, 8, 12–14,
208nn15–16, 18–21
Rosenzweig, Franz, 177, 180, 188n14
Rudner, Jacques, 190
Russell, Bertrand, 33, 42, 49n10, 64,
210
Ryle, Gilbert, 62n1, 187n8

Sagi, Avi, 187n10
Salis, Pietro, 7
Sampson, Riva L., ix
Sani, Filippo, 7
Schrödinger, Erwin, 103
Schwartz, Robert, 208n21
Schwartz, Stephen P., 79n21
scientism, 1, 4, 7n2, 60, 78, 207n4
Searle, John R., 31, 34
Sellars, Wilfrid, xi, 4, 161
semantic indeterminacy, 5, 199,
Sereni, Andrea, 50n18
Shagrir, Oron, 60
skepticism, 30, 95–96, 111–12, 120,
121n3, 150–52, 155, 159n24;
internal, 152, 155

Skinner, B. Frederic, 131
Skolem, Thoralf A., 104–106, 111–12,
117
Skolemization, 103, 106, 111–12, 132
Smart, John J. C., xii, 62
social division of linguistic labor, 67–
68, 78n14
solipsism, 127
Soysal, Zeynep, 122n15
Stegmüller, Wolfgang, 209
Steinitz, Rebecca, 7
stereotype, 23n6, 71, 73
Strawson, Peter, 10–11, 17, 23n3
synonymy, 23n2, 36, 56, 71, 74

Tarski, Alfred, 109, 158n18, 196–97
terms: natural kind, 23n6, 64, 67–73,
75, 76n2, 78n17, 90, 118, 123n21;
observational, 16, 83–85, 87–88,
109, 170; physical magnitude,
23n6, 64, 68–71, 73, 75, 76n2, 90;
theoretical, 17, 63–64, 83–85, 87–88,
90, 92, 98, 170
thick ethical concepts, 174, 187n8
Thorpe, Joshua R., 120, 123n22, 153,
159n24
Timmons, Mark, 188n18
Travis, Charles, 23nn1–7, 157n5,
207n8
truth, 39, 85–88, 92n3, 94, 98–99,
101–3, 110–11, 117, 126, 128, 134–
42, 145–50, 156n2, 158nn16–19,
159nn20–23, 174, 190, 195–97, 204,
206n3, 207nn13–14
Turing, Alan, 53–54
Turing machine, 49n12, 53–55, 59–60
Twin Earth thought experiment, 65–66,
72, 75, 77nn10–11
Tymoczko, Thomas, 124n24

Uehling, Theodore Edward, 7n1
underdetermination, 32
understanding, 2, 64, 76n4, 126–28,
132–41, 146, 150, 155, 203
unrevisability, 21–22, 27, 73

Valore, Paolo, 7
Van Fraassen, Bas C., 122n15
Vega Encabo, Jesús, 207n9
verificationism, x–xi, xiii–xiv, 43, 104,
    111–12, 125–27, 134–36, 140, 207n4
Vineberg, Susan, 50n18
Volpe, Giorgio, 7
Voltolini, Alberto, 7

Wagner, Henri, 78n14
Wagner, Steven J., 49n16
Walsh, Vivian, 171
Wang, Hao, x
West, Cornel, 179, 181

Wettstein, Howard K., 7n1
White, Stephen L., 7
Whitehead, Alfred N., 33, 49n10, 173
Williams, Bernard, 187n8, 199, 207n8
Williams, Michael, 156n2
Wittgenstein, Ludwig, 8n6, 67, 116–17,
    123n23, 128, 146, 149, 154–55,
    158nn16–19, 159n25, 177, 207nn4,
    8, 12, 211
Wright, Crispin, 120, 122n15, 123n22,
    124n24, 138, 152–53, 159n22
Wyatt, Jeremy, 159n22

Zermelo, Ernst, xi, 26, 42, 104

About the Author

Massimo Dell'Utri is Full Professor of Philosophy of Language at the University of Sassari, Italy, Department of Humanities and Social Sciences. He teaches courses in philosophy of language, philosophy of multiculturality, and philosophy of literature and the arts. He has served as the president of the Italian Society for Analytic Philosophy (2018–2021), and he currently serves as the editor-in-chief of the journal *Argumenta* (https://www.argumenta.org/), as member of the European Cultural Parliament and of the Italian Society for the Philosophy of Language. He has written four books (in Italian) and many articles on realism, the concept of truth, and Hilary Putnam's thought (both in English and Italian). His research deals with topics in epistemology, metaphysics, and the philosophy of language, especially the concept of truth.